Aa

a·bi·o·tic fac·tor (ā bī ot′ ik fak′ tər) *noun* The part of an ecosystem that is not alive. See also *biotic factor*. *Water, air, and soil are examples of abiotic factors.*

ab·so·lute age (ab′ sə lüt āj) *noun* The actual age of a rock or fossil formation. See also *relative age*.

ab·so·lute mag·ni·tude (ab′ sə lüt mag′ nə tüd) *noun* The actual brightness of a star. See also *apparent magnitude*.

ab·sorb (əb zôrb′) *verb* To take in. *A plant absorbs water and nutrients from the soil through its roots.* **absorption** *noun*

a·buse (ə būz′) *verb* To use a substance in a way that is harmful to oneself, or to hurt or mistreat people or animals. **abuse** *noun*

a·bys·sal plain (ə bis′ əl plān) *noun* The flattest part of the ocean floor, starting at the base of the continental slope.

ac·cel·er·ate (ak sel′ ə rāt) *verb* To increase speed or to change direction. See also *decelerate*. *A car will accelerate as you press down on the gas pedal.* **acceleration** *noun*

ac·id (as′ id) *noun* A chemical that gives up hydrogen ions (H+) in water, making the water taste sour and giving it a pH below 7. See also *base*. *Lemon juice and vinegar are acids that can be found in many households.* **acid, acidic** *adjectives*

ac·id rain (as′ id rā___ Rain that has a pH lower than 7 ___ pollutants dissolve___

a·cous·tics (ə kü′ s___ The study of sound ___ **acoustic** *adjective*

ac·tion force (ak′ shən fôrs) *noun* In an action-reaction pair, the force that the first object exerts on the second object. See also *reaction force*. *When you step on a stair, your foot exerts an action force on the stair.*

ac·tion-re·ac·tion pair (ak′ shən rē′ ak shən pâr) *noun* A pair of forces that act according to Newton's Third Law of Motion.

ac·tive trans·port (ak′ tiv trans′ pôrt) *noun* The use of energy to move molecules across a cell membrane. See also *passive transport*.

ad·ap·ta·tion (ad ap tā′ shən) *noun* A structure or behavior that helps an organism survive in its environment. *The long neck of a giraffe is an adaptation that helps it to feed on the leaves of tall trees.* **adapt** *verb*

ad·dic·tion (ə dik′ shən) *noun* A very strong need to keep using a substance, even if it causes harm. **addicted, addictive** *adjectives*

ad·he·sive (ad hē′ siv) *adjective* Describes a property of matter in which one kind of matter sticks to other kinds of matter. *Adhesive tape sticks to paper.* **adhesive, adhesion** *nouns*

ad·o·les·cence (ad əl es′ əns) *noun* The period of a life span during which a child develops into an adult. **adolescent** *noun*

aer·o·bic ex·er·cise (ə rō′ bik ek′ sər sīz) *noun* Exercise that causes the body to take in and use oxygen, which makes the heart and lungs stronger.

aer·o·nau·tics (âr ə nôt′ iks) *noun, singular* The science and practice of designing, building, and flying aircraft.

aer·o·sol (âr′ ə sol) *adjective* Describes a liquid that is turned into a fine mist by spraying it with a pressurized gas.

af·ter·shock (af′ tər shok) *noun* A second earthquake that happens shortly after the first, main earthquake.

ai·le·ron (ā′ lə ron) *noun* One of two flaps on an aircraft's wings that control the side-to-side movement of the aircraft (its roll).

air (âr) *noun* The mixture of gases that makes up Earth's atmosphere.

air mass (âr mas) *noun* A huge body of air that has the same temperature, pressure, and humidity throughout.

air pol·lu·tion (âr pə lü′ shən) *noun* Gases and solids in the air that are not normally there, which make the air less healthy for living things.

air pres·sure (âr presh′ ər) *noun* The force that air puts on objects.

air re·sist·ance (âr ri zis′ təns) *noun* The force, also called drag, that air puts on objects as they move through it.

air sac (âr sak) *noun* Part of a bird's respiratory system, a thin-walled sac or bag that connects to the lungs.

al·bi·no (al bī′ nō) *adjective* Describes a person or animal that does not have any pigment (coloring) in the skin, eyes, or hair. *An albino rabbit is all white with pink eyes.* **albino** *noun*

al·co·hol (al′ kə hol) *noun* A colorless liquid that burns easily and in one form is the substance in beer, wine, and some other beverages that can make a person drunk. See also *ethanol. Alcohol use increases the chance of an accident because it slows down a person's reaction time.*

al·gae (al′ jē) *noun, plural* A group of plantlike organisms that live in water and contain the green pigment chlorophyll but lack stems, roots, leaves, or flowers. **alga** *singular*

al·ler·gy (al′ ər jē) *noun* A condition in which a person's immune system reacts to a substance that is harmless to most other people. **allergic** *adjective*

al·loy (al′ oi) *noun* A mixture of two or more different metals or of a metal with another substance.

al·ter·nat·ing cur·rent (ôl′ tər nā tiŋ kûr′ ənt) *noun* A flow of electricity in a circuit that constantly changes direction. See also *direct current. Alternating current in the United States changes direction 60 times every second.* **Abbreviation: AC**

a	cat	e	net	îr	gear	u	cup	u̇	look, pull	*th*	this	ə	alive,
ā	day, lake	ē	seed	o	hot	ū	fuse	oi	soil		hw	wheel		comet,
ä	father	i	fit	ō	cold	ûr	fur, bird	ou	out		zh	measure		acid, atom,
âr	dare	ī	pine	ô	paw	ü	tool, rule	th	thin		ŋ	wing		focus

al·ter·nat·ing gen·er·a·tion (ôl′ tər nā tiŋ jen ə rā′ shən) *noun* A life cycle in which two different forms of the same plant or animal occur every other generation.

al·ter·na·tive en·er·gy source (ôl tûrn′ ə tiv en′ ər jē sôrs) *noun* An energy source, such as the wind or Sun, that is not a fossil fuel.

a·lu·mi·num (ə lü′ mə nəm) *noun* An element that is a low-density, silvery metal. *Aluminum conducts electricity well, resists weathering, and is the most plentiful metal in Earth's crust.* **Symbol: Al**

al·ve·o·li (al vē ō′ lē) *noun, plural* Tiny sacs in the lungs that contain blood vessels where carbon dioxide and oxygen are exchanged. **alveolus** *singular*

a·mi·no ac·ids (ə mē′ nō as′ idz) *noun, plural* Chemicals in living things that are the building blocks of proteins.

am·me·ter (am′ mē tər) *noun* An instrument that measures electric current.

am·mo·ni·a (ə mōn′ yə) *noun* A gas compound made of nitrogen and hydrogen that dissolves easily in water, forming a basic solution. *Many window cleaners contain ammonia.* **Formula: NH_4**

a·moe·ba (ə mē′ bə) *noun* A one-celled animal that lives in water.

am·pere (am′ pîr) *noun* A unit of electric current. **Abbreviation: A**

am·phib·i·an (am fib′ ē ən) *noun* An animal that starts life in water and must return to water to lay eggs but that breathes air using lungs. *Frogs are common amphibians.* **amphibious** *adjective*

am·pli·fy (am′ plə fī) *verb* To make louder.

am·pli·tude (am′ pli tüd) *noun* The height of a wave above its resting position.

a·ne·mi·a (ə nē′ mē ə) *noun* A condition in which a person does not have enough red blood cells or in which the blood cells cannot transport oxygen. **anemic** *adjective*

an·e·mom·e·ter (an ə mom′ i tər) *noun* An instrument used to measure wind speed.

an·gi·o·sperm (an′ jē ə spûrm) *noun* A plant that makes flowers and develops seeds inside an ovary. *Apple trees and roses are examples of angiosperms.*

an·i·mal king·dom (an′ ə məl kiŋ′ dəm) *noun* One of the largest groupings of living things, made up of organisms such as mammals, birds, insects, fish, reptiles, and amphibians.

an·nu·al (an′ ū əl) *adjective* Taking place once a year. *Trees in seasonal climates form annual growth rings.*

an·ther (an′ thər) *noun* The male part of a flower, which makes pollen.

an·thra·cite (an′ thrə sīt) *noun* A type of coal that is hard, black, and shiny.

an·ti·bi·ot·ic (an tē bī ot′ ik) *noun* A medicine that can kill or slow the growth of bacteria. **antibiotic** *adjective*

an·ti·bod·y (an′ ti bod ē) *noun* A substance in the blood that can fight a specific disease. *Measles antibodies in your blood will help you fight the disease.* **antibodies** *plural*

an·ti·gen (an′ ti jən) *noun* A substance that the immune system responds to. *Viruses, bacteria, and pollen are common antigens.*

a·or·ta (ā ôr′ tə) *noun*
The main artery that
carries blood from the
heart to the rest of the
body, except to the
lungs.

ap·par·ent mag·ni·tude
(ə par′ ənt mag′ ni tüd) *noun* How bright
a star or other object appears to be when
seen from Earth. See also *absolute
magnitude*.

aq·ua·cul·ture (äk′ wə kul chər) *noun*
The practice of raising fish or other
marine organisms in pens rather than
catching wild ones.

a·quar·i·um (ə kwâr′ ē əm)
noun A tank set up as an
environment in which
aquatic plants and animals
will live.

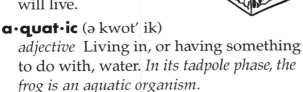

a·quat·ic (ə kwot′ ik)
adjective Living in, or having something
to do with, water. *In its tadpole phase, the
frog is an aquatic organism.*

aq·ui·fer (ak′ wə fər, ä′ kwe fər) *noun*
An underground rock formation that
contains groundwater in its pores and
cracks. *An artesian well drilled into an
aquifer will supply water for a home.*

ar·chae·o·lo·gy (är kē ôl′ ə jē) *noun*
The study of past cultures by observing
the ruins and objects (artifacts) that its
people left behind.

ar·e·a (âr′ ē ə) *noun* The amount of space,
in square units, that a surface takes up.
*They measured the area of the floor in square
meters by multiplying its length in meters by
its width in meters.*

ar·ter·y (är′ tə rē) *noun* A vessel that
carries blood away from the heart toward
other parts of the body.

ar·thro·pod (är′ thrə pod) *noun* A large
group of animals that have jointed legs
and segmented bodies. *Insects, spiders,
scorpions, and crabs are all examples of
arthropods.*

ar·ti·fact (är′ tə fakt) *noun* An object made
and left behind by people who lived long
ago. *Ancient Mayan artifacts are found in
Mexico.*

ar·ti·fi·cial sat·el·lite (är′ tə fish əl
sat′ ə līt) *noun* An object orbiting Earth
that was sent up by a rocket.

a·sex·u·al re·pro·duc·tion
(ā sek′ shü əl rē prə duk′ shən) *noun*
The creation of offspring from just one
parent instead of two. See also *sexual
reproduction.*

as·ter·oid (as′ tə roid) *noun* A mostly
rocky object that orbits the Sun.

as·then·o·sphere (əs then′ ə sfîr) *noun*
An Earth layer made of rock material that
is very hot but still solid and flows slowly.

asth·ma (az′ mə) *noun* A disease in which
a person sometimes has trouble breathing.

as·tro·labe (as′ trə lāb) *noun*
An instrument used to measure the height
of objects in the sky above the horizon.

---

a	cat	e	net	îr	gear	u	cup	ú	look, pull	*th*	this	ə	alive,
ā	day, lake	ē	seed	o	hot	ū	fuse	oi	soil		hw	wheel		comet,
ä	father	i	fit	ō	cold	ûr	fur, bird	ou	out	zh	measure		acid, atom,
âr	dare	ī	pine	ô	paw	ü	tool, rule	th	thin	ŋ	wing		focus

as·tro·nom·i·cal u·nit (as trə nom′ i kəl ū′ nit) *noun* A unit of measure that equals the average distance of Earth from the Sun, or about 150 million kilometers (93 million miles).

as·tro·no·my (ə stron′ ə mē) *noun* The scientific study of objects in space. **astronomer** *noun* **astronomical** *adjective*

a·sym·met·ri·cal (ā si met′ rik əl) *adjective* Not the same, or not balanced, on both sides of an imaginary line or on all sides around a point. See also *symmetrical*.

at·mos·phere (at′ mə sfîr) *noun* The layer of air that surrounds Earth.

at·om (at′ əm) *noun* The smallest particle of an element that still has all the properties of that element.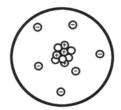

a·tom·ic num·ber (ə tom′ ik num′ bər) *noun* The number of protons in one atom of an element. *Each element has a different atomic number.*

a·tri·a (ā′ trē ə) *noun, plural* Chambers where blood enters the heart, either from the rest of the body or from the lungs. **atrium** *singular*

at·tract (ə trakt′) *verb* To pull toward itself. *A magnet attracts steel and iron objects.* **attraction** *noun*

au·di·o cas·sette (ô′ dē ō kə set′) *noun* A magnetic tape in a plastic case that is used to record and play back sound.

au·di·to·ry nerve (ô′ di tôr ē nûrv) *noun* The fiber that carries information about sound from the ear to the brain.

av·er·age speed (av′ ər ij spēd) *noun* The total distance something travels divided by the time it took to travel. *They went more slowly uphill than downhill, but their average speed for the bike trip was 15 kilometers (9.3 miles) per hour.*

ax·is (ak′ sis) *noun* 1. An imaginary line that a planet spins around. *Earth's axis passes through the North and South Poles.* 2. In mathematics, a horizontal or vertical line at the edge of a graph. **axes** *plural*

ax·on (ak′ son) *noun* A long fiber that branches out from the center of a nerve cell. See also *dendrite*.

back·bone (bak′ bōn) *noun* A long line of bones, called vertebrae, that runs down the back of some animals; also called the spine or spinal column.

back·ground ra·di·a·tion (bak′ ground rā dē ā′ shən) *noun* The amount of radioactivity that is usually present in an area.

bac·ter·i·a (bak tîr′ yə) *noun, plural* One-celled organisms that do not have a nucleus. See also *moneran*. **bacterium** *singular*

bal·anced di·et (bal′ ənst dī′ ət) *noun* Foods eaten regularly that, taken together, contain the amounts of each food type recommended for good health.

bal·anced for·ces (bal' ənst fôrs' əz) *noun, plural* A pair of pushes or pulls that equal each other, work against each other, and add up to zero, canceling each other out. *When balanced forces push on each side of an object, no change in its motion takes place.*

ball-and-sock·et joint (bol ənd sok' it joint) *noun* A place where two bones meet, where the rounded end of one bone (ball) fits into a space in the end of another bone (socket). *The ball-and-socket joint at the shoulder allows the arm to move in a circle.*

bar graph (bär graf) *noun* A diagram that shows data as parallel bars of different heights and is used to compare data.

ba·rom·e·ter (bə rom' i tər) *noun* An instrument used to measure air pressure.

base (bās) *noun* A chemical that gives up hydroxide ions (OH-) in water, making the water taste bitter and giving it a pH above 7. See also *acid. Soap, ammonia, and baking soda are household bases.*
basic *adjective*

bat·ter·y (bat' ə rē) *noun* A container filled with metal plates and chemicals that react to produce electric current. See also *dry cell, wet cell.*

bay (bā) *noun* Part of a coastline that curves inward. *San Francisco Bay is one of the most famous bays in the world.*

Beau·fort scale (bō' fərt skāl) *noun* A set of descriptions used to estimate wind speed, based on how the wind affects objects.

be·hav·ior (bi hāv' yər) *noun* A way that an organism acts.

be·hav·ior·al ad·ap·ta·tion (bi hāv' yər əl ad ap tā' shən) *noun* A way that an organism acts that helps it to survive in its environment. *The behavioral adaptation of "playing dead" helps an opossum escape predators.*

be·hav·ior·al risk fac·tor (bi hāv' yər əl risk fak' tər) *noun* Something that a person does, such as smoking or abusing alcohol, that makes that person more likely to become ill.

bench·mark (bench' märk) *noun* A standard of measurement.

Ber·noul·li prin·ci·ple (bûr nü' lē prin' sə pəl) *noun* The scientific principle that states that the faster a gas or liquid moves, the lower its pressure will be.

bi·ceps (bī' seps) *noun* The large muscle in the upper arm that pulls the lower arm toward the shoulder.

bi·en·ni·al (bī en' ē əl) *adjective* Happening once every two years or taking two years to complete a cycle.
biennial *noun*

big bang the·o·ry (big baŋ thē' ə rē) *noun* The theory that the universe began billions of years ago as a single point that exploded.

a	cat	e	net	îr	gear	u	cup	u̇	look, pull	*th*	this	ə	alive,
ā	day, lake	ē	seed	o	hot	ū	fuse	oi	soil		hw	wheel		comet,
ä	father	i	fit	ō	cold	ûr	fur, bird	ou	out		zh	measure		acid, atom,
âr	dare	ī	pine	ô	paw	ü	tool, rule	*th*	thin		ŋ	wing		focus

bile (bīl) *noun* A fluid made by the liver that helps digest fats.

bi·na·ry star (bī′ nə rē stär) *noun* Two stars that are relatively close together and revolve around each other.

bi·o·di·ver·si·ty (bī ō di vûr′ si tē) *noun* The total number of different species in one area or on Earth as a whole.

bi·o·lo·gi·cal feed·back (bī ə loj′ i kəl fēd′ bak) *noun* The use of the output of body systems to change or control how the system operates.

bi·o·lu·mi·nes·cence (bī ō lü mə nes′ əns) *noun* Light that comes from chemical reactions inside an organism. *A firefly's light comes from bioluminescence.*

bi·o·mass (bī′ ō mas) *noun* Material from living things, usually plants, that is used as a fuel. *Biomass is a renewable energy resource that contains stored energy from sunlight.*

biomass con·ver·sion (bī ō mas kən vûr′ zhən) *noun* The process of breaking down matter that came from living things, such as corn, into a cleaner form of fuel, such as ethyl alcohol.

bi·ome (bī′ ōm) *noun* A large region that has a particular climate and certain types of plants and animals, such as a grassland or rain forest.

bi·o·sphere (bī′ ə sfîr) *noun* All parts of Earth in which organisms can live.

bi·o·tic fac·tor (bī ot′ ik fak′ tər) *noun* The part of an ecosystem that is alive. See also *abiotic factor. Bacteria, fungi, green plants, and animals are examples of biotic factors.*

bi·ot·ic po·ten·tial (bī ot′ ik pə ten′ shəl) *noun* The total number of offspring a species could produce, if nothing in its environment stopped it. *Predators such as owls limit the biotic potential of squirrels in the wild.*

bird (bûrd) *noun* A warm-blooded, egg-laying animal that has a backbone, feathers, and wings.

bi·tu·men (bi tü′ mən) *noun* Tar that occurs naturally and is used to make asphalt.

black hole (blak hōl) *noun* A region in space that is so dense that light cannot escape its gravity.

blad·der (blad′ ər) *noun* A sac structure in an organism that holds a liquid or gas. *The bladder holds urine before it leaves the body.*

block and tack·le (blok ənd tak′ əl) *noun* A simple machine made up of several pulleys strung with the same line.

blood al·co·hol con·cen·tra·tion (blud al′ kə hol kon sən trā′ shən) *noun* A measurement of the amount of alcohol in a person's blood, usually given as a percentage.

blue-green al·gae (blü-grēn al′ jē) *noun, plural* An older term for cyanobacteria.

blue-green bac·ter·i·a (blü-grēn bak tîr′ yə) *noun, plural* An alternate term for cyanobacteria.

boil (boil) *verb* To heat a liquid until it bubbles and evaporates rapidly. *Water boils at 100°C (212°F).*

boil·ing point (boil′ iŋ point) *noun* The temperature at which a liquid boils.

bone (bōn) *noun* A part of a human or animal's skeleton, which is hard and white.

bone mar·row (bōn mâr′ ō) *noun* A soft tissue found in the center of many bones.

Boyle's law (boilz lô) *noun* The scientific law stating that if the temperature and amount of a gas stay the same, then its volume will decrease as the pressure on it increases.

break·er (brā′ kər) *noun* A wave that reaches water too shallow to travel through, so its top crest "breaks," or curls over into foam. *As the waves grew larger, the breakers formed farther from shore.*

bron·chi·al tube (broŋ′ kē əl tüb) *noun* One of two main branches off the trachea, or windpipe.

bron·chi·ole (broŋ′ kē ōl) *noun* A small branch leading off a bronchial tube and into the lung tissue.

bud·ding (bud′ iŋ) *noun* A form of asexual reproduction in which a growth on an organism breaks off to form a new individual. *Yeast reproduces most often by budding.*

buoy·an·cy (boi′ ən sē) *noun* The tendency of an object to sink or float in water.

buoy·ant force (boi′ ənt fôrs) *noun* The upward push of water on an object. *The buoyant forces of the ocean keep huge ships afloat.*

caf·feine (kaf ēn′) *noun* A chemical found in coffee, tea, chocolate, and some soft drinks that acts as a stimulant.

cal·de·ra (kôl dâr′ ə) *noun* A crater at the top of a volcano.

cal·o·rie (kal′ ə rē) *noun* A unit of measure for heat. Also used to refer to a kilocalorie, or food calorie. *Many people count the calories contained in the food they eat so they will not put on too much weight.*

cam·bi·um (kam′ bē əm) *noun* A thin layer in a plant stem where new growth begins. *Tree rings and new buds both grow out of the cambium layer.*

cam·ou·flage (kam′ ə fläzh) *noun* Protective coloring or markings that help an organism blend in with its background. **camouflage** *verb*

can·cer (kan′ sər) *noun* A disease in which some cells multiply out of control, growing into masses of diseased tissue in an organism.

a	cat	e	net	îr	gear	u	cup	u̇	look, pull	*th*	**this**	ə	alive,
ā	day, lake	ē	seed	o	hot	ū	fuse	oi	soil	hw	**wheel**		comet,
ä	father	i	fit	ō	cold	ûr	fur, bird	ou	**out**	zh	measure		acid, atom,
âr	dare	ī	pine	ô	paw	ü	tool, rule	th	**thin**	ŋ	wing		focus

can·yon (kan′ yən) *noun* A steep-walled river valley cut through layers of rock.

ca·pac·i·ty (kə pas′ i tē) *noun* The amount that a container can hold, in units of volume. *The water in the cup overflowed when we exceeded the cup's capacity of 100 mL.*

cap·il·lar·y (kap′ ə lər ē) *noun* The smallest blood vessel, in which blood exchanges gases, food, and waste with cells.

car·bo·hy·drates (kär bō hī′ drāts) *noun, plural* Substances containing carbon, hydrogen, and oxygen that are abundant in foods such as bread and cereal.

carbon (kär′ bən) *noun* An element that is found in all living things and also as a mineral in nature. *Soft, black graphite and hard, clear diamond are both natural forms of carbon.* **Symbol: C**

carbon cy·cle (kär′ bən sī′ kəl) *noun* The movement of carbon from the environment into living things and back into the environment.

carbon di·ox·ide (kär′ bən dī ok′ sīd) *noun* A compound gas that is exhaled by animals as a byproduct of using food and that is used by plants to make food. **Formula: CO_2**

carbon dioxide–ox·y·gen cy·cle (kär′ bən dī ok′ sīd ok′ sə jen sī kəl) *noun* The movement of carbon dioxide and oxygen among animals, plants, and the nonliving parts of the environment.

carbon mon·ox·ide (kär′ bən mon ok′ sīd) *noun* A poisonous gas that is a compound made of equal parts carbon and oxygen. **Formula: CO**

car·di·ac mus·cle (kär′ dē ak mus′ əl) *noun* A kind of muscle found only in the heart.

car·di·o·vas·cu·lar (kär dē ō vas′ kyə lər) *adjective* Having to do with the heart and lungs.

car·ni·vore (kär′ nə vôr) *noun* An animal that eats other animals as food. **carnivorous** *adjective*

car·ti·lage (kär′ tə lij) *noun* A body tissue that keeps its own shape but is not stiff. *Your nose and ear flaps are made of cartilage.*

car·tog·ra·pher (kär′ tog′ rə fər) *noun* A person whose job is to make maps.

cast (kast) *noun* A fossil made when a hollow shape left by an organism filled in with sediments. See also *mold*. *When she split the shale open, she discovered a cast of a dinosaur footprint.*

cat·a·lyst (kat′ ə list) *noun* A substance that starts or speeds up a chemical reaction. **catalytic** *adjective*

ce·les·ti·al (sə les′ chəl) *adjective* Having to do with outer space or the sky.

cell (sel) *noun* The basic unit of living things.

cell dif·fer·en·ti·a·tion (sel dif ə ren shē ā′ shən) *noun* The way that cells in a developing organism take on specific structures and functions.

cell mem·brane (sel mem′ brān) *noun* The layer that surrounds a cell, holds it together, and controls what enters and leaves it.

cell the·or·y (sel thē′ ə rē) *noun* The concept that the cell is the basic unit of all living things.

cel·lu·lar res·pi·ra·tion (sel′ yə lər res pə rā′ shən) *noun* The process by which cells use oxygen to get energy from food molecules.

cell wall (sel wôl) *noun* The thick, stiff layer that grows around plant cells.

Cel·si·us scale (sel′ sē əs skāl) *noun* A temperature scale based on the freezing and boiling points of water, in which water freezes at 0° and boils at 100°.

ce·men·ta·tion (sē men tā′ shən) *noun* The process by which minerals glue together loose sediments, forming sedimentary rock.

Ce·no·zo·ic er·a (sē nə zō′ ik îr′ ə) *noun* Part of the geologic time scale, from 65 million years ago to the present day.

cen·ti·me·ter (sen′ tə mē tər) *noun* A unit of length in the metric system, equal to 1/100 of a meter.

cer·e·bel·lum (ser ə bel′ əm) *noun* A part of the brain that helps coordinate movement and balance.

cer·e·brum (ser′ ə brum, sə rē′ brəm) *noun* The largest part of the brain, where thinking, language, and muscle control take place, and where information from the senses is processed.

chain re·ac·tion (chān rē ak′ shən) *noun* A nuclear reaction in which the splitting of one atom causes the splitting of others.

chan·nel (chan′ əl) *noun* 1. The deepest part of a stream, riverbed, or harbor. 2. In the ocean, a broad strait, such as the English Channel.

chem·i·cal bond (kem′ i kəl bond) *noun* A force that holds atoms together in a molecule.

chemical change (kem′ i kəl chānj) *noun* A change in matter that results in a different kind of matter and that cannot be reversed. See also *physical change*.

chemical en·er·gy (kem′ i kəl en′ ər jē) *noun* Energy that is stored in the bonds between atoms.

chemical e·qua·tion (kem′ i kəl ē kwā′ zhən) *noun* Symbols that show how matter changes during a chemical reaction of elements, for example: $2H_2O = 2H_2 + O_2$.

chemical for·mu·la (kem′ i kəl fôr′ myə lə) *noun* A short way to show the number and kinds of atoms of elements in a molecule, such as NaCl for sodium chloride, or table salt.

chemical prop·er·ty (kem′ i kəl prop′ ər tē) *noun* A property of a substance that can be used to identify it and that describes how that substance reacts with other substances.

chemical re·ac·tion (kem′ i kəl rē ak′ shən) *noun* A change in matter in which two or more elements or compounds (reactants) combine to form different compounds or elements (products).

chemical sym·bol (kem′ i kəl sim′ bəl) *noun* One or two letters that stand for one atom of an element, such as O for oxygen or Au for gold.

a	cat	e	net	îr	gear	u	cup	ủ	look, pull	th	this	ə	alive,
ā	day, lake	ē	seed	o	hot	ū	fuse	oi	soil	hw	wheel		comet,
ä	father	i	fit	ō	cold	ûr	fur, bird	ou	out	zh	measure		acid, atom,
âr	dare	ī	pine	ô	paw	ü	tool, rule	th	thin	ŋ	wing		focus

chemical weath·er·ing (kem' i kəl weth' ər iŋ) *noun* The wearing away of a rock as it reacts with air, water, and soil. See also *mechanical weathering*.

chem·is·try (kem' is trē) *noun* The study of matter, its properties, and its behavior.

chlo·rine (klôr' ēn) *noun* An element that is a greenish yellow gas, used to purify water and as a disinfectant and bleach. **Symbol: Cl**

chlo·ro·fluo·ro·car·bons (klôr ō flür ō kär' bənz) *noun, plural* Artificial compounds known as CFCs that have been used in air conditioners and aerosol cans and that may be destroying ozone in the upper atmosphere.

chlo·ro·phyll (klôr' ə fil) *noun* The green pigment compound in plants that captures sunlight and makes food out of water and carbon dioxide.

chlo·ro·plast (klôr' ə plast) *noun* The part of a plant cell that contains chlorophyll.

cho·les·ter·ol (kə les' tə rôl) *noun* A fatty substance in the blood and in other body tissues.

chro·mo·some (krō' mə sōm) *noun* A structure in the cell nucleus that contains the genetic material DNA.

chro·mo·sphere (krō' mə sfîr) *noun* The layer of the Sun's atmosphere below the corona that is visible only during a total solar eclipse.

cil·i·a (sil' ē ə) *noun, plural* Tiny hairlike structures on the surface of some cells and one-celled organisms, capable of rhythmic movement. **cilium** *singular*

cin·der-cone vol·ca·no (sin' dər kōn vol kā' nō) *noun* A steep-sided volcano built up from eruptions of rock, cinders, and ash.

cir·cuit (sûr' kit) *noun* A complete path that electric current moves through. *When you turn on a light switch, you complete a circuit and the light goes on.*

circuit break·er (sûr' kit brā' kər) *noun* A device that opens, or breaks, the circuit when there is more current than is safe.

circuit di·a·gram (sûr' kit dī ' ə gram) *noun* A picture that shows how parts of a circuit are connected.

cir·cu·la·tion (sûr kyə lā' shən) *noun* The movement of something through a path that makes a closed loop. *The circulation of blood through the body begins and ends with the heart.* **circulate** *verb*

cir·cu·la·to·ry sys·tem (sûr' kyə lə tôr ē sis' təm) *noun* The organ system made up of the heart and blood vessels that moves blood throughout the body.

cir·rus cloud (sîr' əs kloud) *noun* A wispy cloud made of ice crystals, high in the troposphere.

class (klas) *noun* A grouping of organisms in the classification scheme, more specific than phylum but less specific than order.

clas·si·fi·ca·tion (klas ə fi kā' shən) *noun* A way of organizing things into groups based on ways in which they are similar or related. **classify** *verb*

clav·i·cle (klav' ə kəl) *noun* The collarbone, which connects the shoulders and breastbone.

clay (klā) *noun* A sediment made of tiny flakes of chemically weathered rock.

cleav·age (klē' vij) *noun* A flat surface where a mineral tends to break.

cli·mate (klī′ mət) *noun* The average weather in a particular region over many years.

climate zone (klī′ mət zōn) *noun* A large region that has about the same average temperature and rainfall throughout.

clone (klōn) *verb* To grow a new plant or animal that is identical to one adult parent. **clone** *noun*

closed cir·cuit (klōzd sûr′ kit) *noun* A circuit that has no breaks, allowing an electric current to flow through it. See also *open circuit.*

cloud (kloud) *noun* A cluster of water droplets or ice crystals hanging or floating in the atmosphere.

coal (kōl) *noun* A fossil fuel formed from the remains of plants and animals that were buried deep inside Earth for millions of years.

coast·al o·cean (kōs′ təl ō′ shən) *noun* The part of the ocean that lies near the coast of a continent and is affected by materials coming off the continent.

coch·le·a (kok′ lē ə) *noun* A coiled structure in the inner ear that senses sound waves.

cold-blood·ed (kōld′ blud əd) *adjective* Describes an animal whose body temperature changes with the temperature of its environment. See also *warm-blooded.*

cold front (kōld′ frunt) *noun* The boundary between two air masses, where a colder air mass is pushing a warmer air mass in front of it. See also *warm front.*

col·loid (kol′ oid) *noun* A mixture in which tiny clumps of molecules are spread throughout a liquid and do not settle out on their own. See also *emulsion, suspension.*

col·o·nize (kol′ ə nīz) *verb* To move into a new area and populate it. *Plants and animals may colonize new areas when their habitats become overcrowded.*

col·o·ny (kol′ ə nē) *noun* A group of organisms of the same species that live together and depend on each other, such as a colony of ants.

com·bus·ti·ble (kəm bus′ tə bəl) *adjective* Describes something that can burn. **combustibility** *noun*

com·et (kom′ ət) *noun* An object made of ice and dust that orbits the Sun. *A comet develops a tail as it moves near the Sun.*

com·men·sal·ism (kə men′ sə liz əm) *noun* A relationship between two different species that helps one species and does not help or hurt the other species.

com·mu·ni·ca·ble dis·ease (ke mū′ ni kə bəl di zēz′) *noun* An illness that can be passed from one person to another.

a	cat	e	net	îr	gear	u	cup	u̇	look, pull	*th*	this	ə	alive,
ā	day, lake	ē	seed	o	hot	ū	fuse	oi	soil	hw	wheel		comet,
ä	father	i	fit	ō	cold	ûr	fur, bird	ou	out	zh	measure		acid, atom,
âr	dare	ī	pine	ô	paw	ü	tool, rule	th	thin	ŋ	wing		focus

com·mu·ni·ty (kə mū′ ni tē) *noun*
A group of organisms living together in a certain area. *Deer are common members of a deciduous forest community.*

com·pact disc (kom′ pakt disk) *noun*
A flat, round object used to store music or computer software.

com·pac·tion (kəm pak′ shən) *noun*
The process in which loose sediments buried deep inside Earth are pressed together until they form sedimentary rock.

com·pare (kəm pâr′) *verb* To describe how two things are like each other. See also *contrast.* **comparison** *noun*

com·pass (kum′ pəs) *noun* A device used to find magnetic north. *They used a compass to find their way out of the forest.*

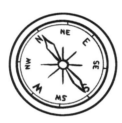

com·pe·ti·tion (kom pə tish′ ən) *noun*
In an ecosystem, the contest among organisms for resources, such as sunlight, water, and food.

com·plete flow·er (kəm plēt′ flou′ ər) *noun* A flower that has all four flower parts: sepals, petals, pistils, and stamens. See also *incomplete flower.*

com·po·site-cone vol·ca·no (kəm poz′ it kōn vol kā′ nō) *noun*
A steep-sided volcano built up from eruptions of both cinders and lava.

com·pound (kom′ pound) *noun* Matter made up of two or more elements that are chemically bonded and having properties that are different from the elements that make it up. *Water is a compound made from oxygen and hydrogen.*

compound ma·chine (kom′ pound mə shēn′) *noun* A tool made up of two or more simple machines that makes work easier.

compound mi·cro·scope (kom′ pound mī′ krə scōp) *noun* A device that uses two lenses to make an object look larger.

compound pul·ley (kom′ pound pŭl′ ē) *noun* A machine made of two or more grooved wheels, or pulleys.

com·pres·sion (kəm presh′ ən) *noun*
The part of a sound wave where particles are pushed close together. See also *rarefaction. The length of a sound wave can be measured from one compression to the next compression.* **compress** *verb*

con·cave lens (kon kāv′ lenz) *noun*
A round, clear piece of glass or plastic that is thinner in the center and thicker at the edges. See also *convex lens. People who are nearsighted often wear glasses with concave lenses.*

concave mir·ror (kon kav′ mîr′ ər) *noun*
A surface that is curved inward like a bowl and that reflects light.

con·cen·tra·tion (kon sen trā′ shən) *noun*
The amount of solute in a solution compared with the amount of solvent. *The concentration of salt in ocean water is about 3.5 percent.*

con·clu·sion (kən clü′ zhən) *noun*
A statement saying whether the results of an experiment support the hypothesis.

con·crete (kon′ krēt) *noun* A mixture of cement, rock, sand, and water that is used to make sidewalks, roads, and buildings.

con·crete (kon krēt′) *adjective* Describes something that is solid and specific. *A well-written science report contains concrete examples that support the main idea.*

con·den·sa·tion (kon den sā′ shən) *noun* Part of the water cycle in which water vapor changes into liquid water. Also, the water drops that form by condensing.

con·dense (kən dens′) *verb* To change from a gas into a liquid.

con·di·tion·ing (kən dish′ ə niŋ) *noun* The process of learning to connect one event with another event that usually occurs with it. *Conditioning prompted the class to expect a quiz every Friday.*

con·duct (kən dukt′) *verb* To transfer energy, such as electricity or heat. See also *insulate.*

con·duc·tion (kən duk′ shən) *noun* The transfer of heat by direct contact.

con·duc·tor (kən duk′ tər) *noun* Something that transfers heat or electricity. See also *insulator.*

con·i·fer (kon′ ə fər) *noun* A plant that bears seeds in cones and has long, thin leaves (needles) that usually stay on all year.

con·ju·ga·tion (kon jə gā′ shən) *noun* The transfer of genes by two cells joining together, which happens in some algae and bacteria.

con·serve (kən sûrv′) *verb* To use a resource such as air, water, or land carefully, allowing it to last longer. **conservation** *noun*

con·stant (kon′ stənt) *noun* A value that stays the same in an equation. See also *variable.*

con·stel·la·tion (kon stə lā′ shən) *noun* A group of stars that appears to make a pattern in the sky.

con·sum·er (kən sü′ mər) *noun* An organism that survives by eating other organisms in an ecosystem.

con·tact lens (kon′ takt lenz) *noun* A curved piece of clear plastic that fits directly on the eye and corrects vision.

con·ta·gious (kən tā′ jəs) *adjective* Describes a disease easily passed from one person to another.

con·tam·i·nant (kən tam′ ə nənt) *noun* An element or compound that, when mixed with another substance, makes that substance impure or unclean. **contaminate** *verb*

con·ti·nen·tal drift (kon tə nen′ təl drift) *noun* An old theory that the continents moved through the ocean crust. *Continental drift was the basis for today's theory of plate tectonics.*

continental edge (kon tə nen′ təl ej) *noun* The end of the continental shelf, where it meets the continental slope.

continental rise (kon tə nen′ təl rīz) *noun* The area between the continental shelf and the abyssal plain, having a slope less steep than the continental slope.

a	cat	e	net	îr	gear	u	cup	ů	look, pull	*th*	this	ə	alive,
ā	day, lake	ē	seed	o	hot	ū	fuse	oi	soil	hw	wheel		comet,
ä	father	i	fit	ō	cold	ûr	fur, bird	ou	out	zh	measure		acid, atom,
âr	dare	ī	pine	ô	paw	ü	tool, rule	th	thin	ŋ	wing		focus

continental shelf (kon tə nen′ təl shelf) *noun* A gently sloping, underwater edge of a continent.

continental slope (kon tə nen′ təl slōp) *noun* The steep slope on the ocean floor at the edge of a continental shelf, which ends in the continental rise.

con·tract (kən trakt′) *verb* To pull together into a smaller space. See also *expand. An inflated balloon will contract when you let the air out of it.*

con·trast (kən trast′) *verb* To describe how two things are different from each other. See also *compare.* **contrast** *noun*

con·trol (kən trōl′) *noun* A condition or factor that is kept the same throughout an experiment. See also *variable. Our controls for the experiment were the amount of water, water temperature, and mass of salt.*

con·vec·tion (kən vek′ shən) *noun* The transfer of heat by currents within a liquid or gas.

convection cell (kən vek′ shən sel) *noun* An up-and-down, circular movement of air or water that is caused by temperature differences.

convection cur·rent (kən vek′ shən kûr′ ənt) *noun* An up or down movement in a liquid or gas that is caused by differences in temperature.

con·verge (kən vûrj′) *verb* To come together. *Light rays converge after passing through a convex lens.* See also *diverge.*

con·ver·gent bound·a·ry (kən vûr′ jənt boun′ də rē) *noun* A place where two crustal plates are moving toward each other.

con·vex lens (kon veks′ lenz) *noun* A round, clear piece of glass or plastic that is thicker in the center and thinner at the edges. See also *concave lens. A convex lens makes things look larger by bending light rays together.*

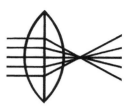

convex mir·ror (kon veks′ mîr′ ər) *noun* A surface that is curved outward and that reflects light.

co·or·di·nate (ko ôr′ dən āt) *verb* To bring together two or more things so they work together. *The two students coordinated their efforts on the project, each taking a different part of the work.*

co·or·di·nates (ko ôr′ dən əts) *noun, plural* A set of two numbers that describes where something is located in space or on a graph.

core (kôr) *noun* The innermost layer of Earth.

Co·ri·o·lis ef·fect (kôr ə ō′ ləs ə fekt′) *noun* The curved path a moving object takes across Earth's surface, caused by Earth turning beneath it.

cor·ne·a (kôr nē ə) *noun* The clear covering over the iris and pupil, in the eye.

co·ro·na (kə rō′ nə) *noun* The outermost layer of the Sun's atmosphere, past the chromosphere, which is visible only during a solar eclipse.

cor·tex (kôr′ teks) *noun* A layer of plant tissue right below the outermost covering on stems and roots.

cos·mo·lo·gy (koz mol′ ə jē) *noun* The study of theories about the origin of the universe.

cot·y·le·don (kot ə lē′ dən) *noun* The seed leaf, which provides food for a young plant while it is sprouting.

cra·ni·um (krā′ nē əm) *noun* The skull.

cra·ter (krā′ tər) *noun*
A bowl-shaped landform made when an object hits a planet's or moon's surface or when a volcano erupts.

crest (krest) *noun* The uppermost peak of a wave. *The length of an ocean wave can be measured from one crest to the next crest.*

cre·vasse (krə vas′) *noun* A deep crack in the ice of a glacier.

crop ro·ta·tion (krop rō tā′ shən) *noun* The planting of different crops in a field each year so the soil does not run out of certain nutrients.

cross·breed (kros′ brēd) *noun* A plant or animal with parents from two different varieties. *Cockapoos are a crossbreed of cocker spaniels and poodles.* **crossbreed** *verb*

cross-pol·li·na·tion (kros pol ə nā′ shən) *noun* The result when pollen from one plant fertilizes the eggs of another plant. See also *self-pollination.*

crust (krust) *noun* The outermost, rocky layer of Earth. *Earth's crust rests on top of a softer layer called the mantle.*

crus·ta·cean (krus tā′ shən) *noun* An animal, such as a crab or shrimp, that has a hard exoskeleton, a pair of modified front legs, and antennae. *Most crustaceans live in water.*

crys·tal (kris′ təl) *noun* A sample of a mineral or chemical whose atoms are arranged in a regular pattern, giving it a certain shape and other physical properties.

cu·mu·lus cloud (kū′ myə ləs kloud) *noun* A puffy fair-weather cloud often with a flat base and "heaped up" top.

cur·rent (kûr′ ənt) *noun* A flow of water, air, or electricity.

cy·a·no·bac·te·ri·a (sī an ō bak tîr′ yə) *noun, plural* One-celled organisms that do not have a nucleus, live in water, and can make their own food; also called blue-green bacteria.

cy·cad (sī′ kad) *noun* A kind of tropical plant that looks something like a palm but belongs to a different order. *Cycads were most common during the age of the dinosaurs.*

cy·cle (sī′ kəl) *noun* A series of events that occurs over and over in a certain order.

cy·clone (sī′ klōn) *noun* A storm, such as a tornado, that whirls around a low-pressure center.

cy·to·plasm (sī′ tō plaz əm) *noun* The material that fills a cell, outside the cell nucleus.

a	cat	e	net	îr	gear	u	cup	u̇	look, pull	*th*	**this**	ə	alive,
ā	day, lake	ē	seed	o	hot	ū	fuse	oi	soil	hw	**wheel**		comet,
ä	father	i	fit	ō	cold	ûr	fur, bird	ou	**out**	zh	measure		acid, atom,
âr	dare	ī	pine	ô	paw	ü	tool, rule	th	**thin**	ŋ	wing		focus

da·ta (dat′ ə, dā′ tə) *noun, plural*
Information gathered during an
experiment. **datum** *singular*

de·cel·er·a·te (dē sel′ ə rāt) *verb*
To change speed by slowing down. See
also *accelerate.* **deceleration** *noun*

dec·i·bel (des′ ə bəl) *noun* A unit of
measure for the loudness, or volume, of
sound. **Abbreviation: dB**

de·cid·u·ous (di sij′ ü əs) *adjective*
Describes an organism that drops
something during part of its life cycle.
Deciduous trees lose their leaves each autumn.

deciduous for·est (di sij′ ü əs fôr′ əst)
noun A forest biome based on trees that
lose their leaves each autumn.

de·com·pos·er (dē kom pōz′ ər) *noun*
An organism, such as a bacterium or
fungus, that breaks down dead matter,
returning it to the soil. **decompose** *verb*

de·com·po·si·tion (dē kom pə zish′ ən)
noun 1. A reaction in which a compound
breaks down into the elements that make
it up. 2. The act of breaking down into
simpler parts.

de·fi·cient (di fish′ ənt) *adjective* Lacking
in something that is needed. *Illness can
result when a person's diet is deficient in basic
food groups.*

de·for·es·ta·tion (dē fôr əs tā′ shən) *noun*
Removing all the trees from an area.

deg·ra·da·tion (deg rə dā′ shən) *noun*
The wearing away of a surface by erosion.
degrade *verb*

de·gree (di grē′) *noun* 1. A unit on a
temperature scale. *The temperature was 3
degrees Celsius.* 2. A unit of measure for an
angle or arc. *A right angle equals 90 degrees.*
Symbol: °

del·ta (del′ tə) *noun* A deposit of sediments
where a stream enters a larger, calmer
body of water.

den·drite (den′ drīt) *noun* A short fiber
with many branches that is attached to the
center of a nerve cell. See also *axon.*

de·nom·i·na·tor (di nom′ ə nā tər) *noun*
The part of a fraction that is below the
line. See also *numerator.*

den·si·ty (den′ si tē) *noun* A physical
property that describes the mass per unit
volume of a substance. *The density of iron
is greater than the density of water, so iron
will sink in water.*

density cur·rent (den′ si tē kûr′ ənt) *noun*
A movement in a body of water or air that
happens because of differences in density
within the water.

de·pen·dence (di pen′ dəns) *noun*
1. A strong need for a drug such as
alcohol or cocaine, which occurs after a
person has used the drug. 2. A reliance
on a substance or organism for survival.

de·ple·tion (di plē′ shən) *noun* The loss of
certain minerals or nutrients from the soil.

de·po·si·tion (dep ə zish′ ən) *noun*
The dropping of sediments by water,
wind, or melting ice onto a solid surface.
See also *erosion.*

der·mis (dûr′ məs) *noun* The thick inner
layer of the skin. See also *epidermis.*

de·sal·i·na·tion (dē sal ə nā' shən) *noun* The removal of salt from sea water so it can be used as fresh water.

des·ert (dez' ərt) *noun* An area that gets less than 25 centimeters (about 10 inches) of precipitation each year, making it very dry. *The Sahara Desert gets very little rain each year.*

dew (dü) *noun* Water droplets that form when the temperature of humid air drops.

dew point (dü point) *noun* Temperature at which dew starts to form on surfaces. *Dew point depends on both temperature and humidity.*

di·a·phragm (dī' ə fram) *noun* A large muscle between the chest and the abdomen that is involved in breathing.

di·a·tom (dī' ə tom) *noun* A microscopic shelled organism that lives in ocean water.

di·cot (dī' kot) *noun* Short for dicotyledon, a plant, such as a bean, that has two seed leaves. See also *monocot, seed leaf.*

di·et·ar·y fat (dī' i ter ē fat) *noun* Fats or oils that are eaten on a regular basis.

dif·fu·sion (di fū' zhən) *noun* Movement of a substance from a place where there is more of it to a place where there is less of it. *As diffusion brought the smell of baking bread from the kitchen into the living room, we began to feel hungry.* **diffuse** *verb*

di·ges·tion (dī jes' chən) *noun* The process of breaking down food into molecules that can be used by cells. **digest** *verb*

di·ges·tive sys·tem (dī jes' tiv sis' təm) *noun* A group of organs that work together to break food down so it can be used by the body.

dig·it (dij' it) *noun* 1. A single finger or toe. *The human foot has five digits.* 2. A single place value in a larger number. *The number 1,965 contains the digits 1, 9, 6, and 5.*

dike (dīk) *noun* A mass of rock that formed when magma flowed into a crack in another rock, then cooled.

di·no·saurs (dī' nə sôrz) *noun, plural* A group of extinct organisms, with some similarities to reptiles, that lived from 250 million to 65 million years ago.

dip·loid cell (dīp' loid sel) *noun* A cell that has two sets of chromosomes, one from each parent. See also *haploid cell.*

di·rect cur·rent (di rekt' kûr' ənt) *noun* A flow of electricity that moves in one direction. See also *alternating current. Dry cells provide direct current.*
Abbreviation: DC

direct de·vel·op·ment (di rekt' di vel' əp mənt) *noun* In contrast to metamorphosis, a life cycle that does not involve significant changes in form.

a	cat	e	net	îr	gear	u	cup	u̇	look, pull	*th* **this**	ə	alive,
ā	day, lake	ē	seed	o	hot	ū	fuse	oi	soil	hw **wheel**		comet,
ä	father	i	fit	ō	cold	ûr	fur, bird	ou	out	zh **measure**		acid, atom,
âr	dare	ī	pine	ô	paw	ü	tool, rule	th	thin	ŋ **wing**		focus

dis·charge (dis' chärj) *noun* 1. The water flowing through a stream at a particular point. *The discharge of streams and rivers increases when winter snows melt.* 2. The sudden movement of an electric charge from a point of buildup to another point. *Lightning is an electrical discharge between clouds and the ground.*

dis·ease (di zēz') *noun* Illness or sickness of a plant or animal.

dis·solve (di zôlv') *verb* To spread evenly throughout another substance, especially a solid spreading out through a liquid. *As she swirled the beaker, the salt dissolved into the water.*

dis·tance (dis' təns) *noun* How far it is from one point to another.

dis·till (di stil') *verb* To separate a solution by first evaporating the liquid part, then condensing it. **distillation** *noun*

di·verge (dī vûrj') *verb* To move apart. *Light rays diverge after passing through a concave lens.* See also *converge.*

di·ver·gent bound·a·ry (dī vûr' jənt boun' də rē) *noun* A place where two of Earth's crustal plates are moving away from each other.

di·ver·si·ty (di vûr' si tē) *noun* The state of having many different kinds of organisms or genes. **diverse** *adjective*

DNA (dē en ā) *noun* Short for deoxyribonucleic acid, the chemical that holds genetic instructions for growing and maintaining an organism.

dome moun·tain (dōm moun' tən) *noun* A large, rounded mass of igneous rock that cooled underground, then was exposed by erosion.

do·mes·ti·cat·ed (də mes' ti kāt əd) *adjective* An animal or plant that has lived with humans and been bred by them for many generations.

dom·i·nant gene (dom' i nənt jēn) *noun* In a pair of genes, the one that determines a trait in an organism if it is present. See also *recessive gene.*

dominant trait (dom' i nənt trāt) *noun* The trait that an organism has if the dominant gene for that trait is present. See also *recessive trait.*

Dop·pler ef·fect (dop' lər ə fekt') *noun* The change in wave frequency that seems to happen when either the wave source or its observer is moving. *A siren sounds higher, then lower, as it passes you because of the Doppler effect.*

dor·mant (dôr' mənt) *adjective* Describes something that is not active or growing now but will be. *Seeds are dormant until the temperature and moisture are right for germination.*

down·draft (doun' draft) *noun* A strong wind that blows down toward the ground.

drag (drag) *noun* A force that pushes against an object that is moving through air or water. *The shape of a jet plane reduces drag so the plane can move more quickly through the air.*

dry cell (drī sel) *noun* An electrochemical cell, sometimes called a battery, that uses a moist paste of chemicals to provide electrical energy to a circuit.

dy·nam·ic e·qui·lib·ri·um (dī nam′ ik ē kwə lib′ rē əm) *noun* Two processes or reactions happening at the same rate so that it looks as if nothing is changing, when in fact change is occurring.

ear·drum (îr′ drum) *noun* A thin membrane in the ear that vibrates when sound waves hit it.

Earth (ûrth) *noun* The planet we live on; the third planet from the Sun in our solar system.

earth·quake (ûrth′ kwāk) *noun* A sudden movement of the ground that happens when part of Earth's crust suddenly shifts, usually at a fault line.

e·chi·no·derm (ə kī′ nə dûrm) *noun* An animal that lives in the ocean, has a firm, spiny skin, and is symmetrical around a point. *Starfish and sea urchins are examples of echinoderms.*

ech·o (ek′ ō) *noun* A sound wave that reflects off a surface and back to its source. **echo** *verb*

ech·o·lo·ca·tion (ek ō lō kā′ shən) *noun* Finding things by sending out sound waves that reflect off objects.

e·clipse (ē klips′) *noun* The blocking of one object in space by another. *In a solar eclipse, the Moon blocks the Sun from Earth, and in a lunar eclipse, Earth blocks the Moon from the Sun.*

ec·o·log·i·cal suc·ces·sion (ek ə loj′ i kəl sək sesh′ ən) *noun* A natural change in the community of plants and animals that are dominant in an area.

e·col·o·gy (ē kol′ ə jē) *noun* The study of how living things interact with one another and with their environment.

e·co·sys·tem (ek′ ō sis təm) *noun* All the living and nonliving things in an area and their interactions with one another.

ef·fi·cien·cy (i fish′ ən sē) *noun* A measure of the amount of work it takes to use a machine compared with the amount of work the machine produces.

ef·fort dis·tance (ef′ ərt dis′ təns) *noun* The distance that the effort force moves a machine. *The effort distance is greater than the load distance in most simple machines.*

effort force (ef′ ərt fôrs) *noun* The amount of force that is put into a machine. *The effort force is less than the load force in most simple machines.*

a	cat	e	net	îr	gear	u	cup	u̇	look, pull	*th*	**this**	ə	alive,
ā	day, lake	ē	seed	o	hot	ū	fuse	oi	soil	hw	**wheel**		comet,
ä	father	i	fit	ō	cold	ûr	**fur**, bird	ou	**out**	zh	measure		acid, atom,
âr	dare	ī	pine	ô	paw	ü	**tool**, rule	th	**thin**	ŋ	wing		focus

egg (eg) *noun* 1. A female reproductive cell. 2. A fertilized reproductive cell with a shell that will develop into an animal, such as a bird or reptile.

e·las·tic·i·ty (i la stis′ i tē) *noun* The ability of some kinds of matter to bend or stretch and then return to their original shape. *The elasticity of rubber is what lets a rubber band snap back after being stretched.*

e·lec·tri·cal cir·cuit (i lek′ tri kəl sûr′ kit) *noun* A complete path that electric current moves through.

electrical en·er·gy (i lek′ tri kəl en′ ər jē) *noun* An ability to do work (energy) that comes from the movement of electric charges.

e·lec·tric charge (i lek′ trik chärj) *noun* A property of matter that comes from particles within the atom; electrons have a negative charge and protons have a positive charge.

electric cur·rent (i lek′ trik kûr′ ənt) *noun* The movement of electric charges in a conductor.

electric force (i lek′ trik fôrs) *noun* The attraction or repulsion of two objects because of their electric charges.

e·lec·tric·i·ty (i lek tris′ i tē) *noun* A general term describing force, energy, current, or power that comes from the movement of electric charges.

e·lec·tro·chem·i·cal cell (i lek trō kem′ i kəl sel) *noun* A device that uses a chemical reaction to provide electrical energy to a circuit. See also *battery, dry cell, wet cell.*

e·lec·trode (i lek′ trōd) *noun* A rod that carries electric current into a solution.

e·lec·tro·mag·net (i lek′ trō mag′ nit) *noun* A steel bar with a wire wrapped around it that becomes a magnet when electric current moves through the wire and loses its magnetism when the current is stopped.

e·lec·tro·mag·net·ic ra·di·a·tion (i lek trō mag net′ ik rā dē ā′ shən) *noun* Energy in the form of waves that can travel through empty space. *Light, radio waves, and ultraviolet are some kinds of electromagnetic radiation.*

electromagnetic spec·trum (i lek trō mag net′ ik spek′ trəm) *noun* All the frequencies of electromagnetic radiation, listed in order.

electromagnetic wave (i lek trō mag net′ ik wāv) *noun* Any wave that is part of the electromagnetic spectrum.

e·lec·tro·mag·ne·tism (i lek trō mag′ nə tiz əm) *noun* The study of the relationship between magnetism and electricity.

e·lec·tron (i lek′ tron) *noun* A negatively charged particle that is the smallest of the particles that make up an atom.

electron mi·cro·scope (i lek′ tron mī′ krə skōp) *noun* A device that uses a beam of electrons to view an object and make it appear larger.

el·e·ment (el′ ə mənt) *noun* A form of matter that is made up of only one kind of atom. *Nitrogen is the most common element in Earth's atmosphere.*

el·e·va·tion (el ə vā′ shən) *noun* The height of an object or place measured from average sea level.

el·lip·ti·cal (i lip′ ti kəl) *adjective* Having an oval shape with two focuses. *Johannes Kepler figured out that planetary orbits around the Sun are elliptical rather than circular.*

elliptical gal·ax·y (i lip′ ti kəl gal′ ək sē) *noun* A huge, oval-shaped group of stars that does not have a spiral structure.

El Ni·ño (el nēn′ yo) *noun* An unusually warm ocean current in the eastern Pacific Ocean near the equator that causes shifts in weather patterns in other parts of the world as well. See also *La Niña*.

em·bry·o (em′ brē ō) *noun* An animal or plant in its very early stages of development.

em·phy·se·ma (em fi sē′ mə) *noun* A condition in which the air sacs (alveoli) in a person's lungs are enlarged, making it difficult to breathe.

e·mul·sion (i mul′ shən) *noun* A mixture in which small droplets of one liquid are spread throughout another liquid. See also *colloid, suspension*.

en·dan·gered spe·cies (en dān′ jərd spē′ shēz) *noun* A kind of plant or animal that is at risk of extinction. *The humpback whale is an example of an endangered species.*

en·do·crine sys·tem (en′ də krin sis′ təm) *noun* A group of organs that give instructions to the body through chemicals called hormones.

en·do·skel·e·ton (en dō skel′ i tən) *noun* A hard, jointed structure inside an animal, usually made of bone. See also *exoskeleton*. *Birds, fish, and mammals have endoskeletons.*

en·do·sperm (en′ dō spûrm) *noun* Food that is stored in a seed for the plant to use while sprouting.

en·do·ther·mic (en dō thûr′ mik) *adjective* Describes a chemical reaction that takes in heat energy.

en·er·gy (en′ ər jē) *noun* The ability to do work. *A healthy breakfast provided him with the energy to ride his bicycle all morning.*

energy pyr·a·mid (en′ ər jē pîr′ ə mid) *noun* A diagram that shows how energy is used up as it travels through food chains in an ecosystem.

en·gine (en′ jin) *noun* A device that changes stored energy (potential energy) into motion (kinetic energy).

en·vi·ron·ment (en vī′ rən mənt) *noun* All the living and nonliving things that surround an organism where it lives.

en·vi·ron·men·tal risk fac·tor (en vī rən mən′ təl risk fak′ tər) *noun* Something in a person's environment, such as air pollution, that makes that person more likely to become ill.

a	cat	e	net	îr	gear	u	cup	ù	look, pull	*th*	this	ə	alive,
ā	day, lake	ē	seed	o	hot	ū	fuse	oi	soil		hw	wheel		comet,
ä	father	i	fit	ō	cold	ûr	fur, bird	ou	out	zh	measure		acid, atom,
âr	dare	ī	pine	ô	paw	ü	tool, rule	*th*	thin	ŋ	wing		focus

en·zyme (en′ zīm) *noun* A protein made by a living thing that speeds up certain chemical reactions.

ep·i·cen·ter (ep′ i sen tər) *noun* The point on Earth's surface directly above an earthquake's focus.

ep·i·der·mis (ep i dûr mis) *noun* The outermost layer of skin. See also *dermis.*

ep·i·glot·tis (ep i glot′ is) *noun* A flap of cartilage that keeps food out of the windpipe by covering it during swallowing.

e·qua·tor (i kwā′ tər) *noun* An imaginary line around the middle of a planet or moon halfway between its north and south poles.

e·qui·li·bri·um (ē kwə lib′ rē əm) *noun* A situation in which forces or reactions are balanced and no overall change is taking place.

er·a (îr′ ə) *noun* The second-largest division of time in the geological time scale, equal to tens or hundreds of millions of years.

e·ro·sion (i rō′ zhən) *noun* The wearing away of soil and rock by water, wind, or ice. See also *deposition.*

e·soph·a·gus (i sof′ ə gəs) *noun* The muscular tube that connects the throat with the stomach.

es·ti·mate (es′ ti māt) *verb* To make a reasonable guess at a number value, based on some information. **estimate** *noun*

es·tu·ar·y (es′ chü er ē) *noun* A place where a river or stream meets an ocean and fresh water mixes with salt water.

eth·a·nol (eth′ ə nol) *noun* The kind of alcohol that is in alcoholic beverages, such as beer and wine.

eu·gle·na (ū glē′ nə) *noun* A type of one-celled organism that lives in fresh water and can make its own food.

e·vap·o·rate (i vap′ ə rāt) *verb* To change from a liquid into a gas.

e·vap·o·ra·tion (i vap ə rā′ shən) *noun* The part of the water cycle where liquid water changes into water vapor.

ev·er·green (ev′ ər grēn) *noun* A plant that has green leaves or needles all year, such as a pine tree or rhododendron bush.

ev·o·lu·tion (ev ə lü′ shən) *noun* The theory that species gradually change over time, through random variations that improve the chances of certain individuals to survive and reproduce. *The evolution of the horse from a small, doglike ancestor can be observed in the fossil record.* See also *natural selection.* **evolve** *verb*

ex·ca·vate (eks′ kə vāt) *verb* To dig a hole in something. *The workers excavated the dirt and rock so they could build the foundation for a house.*

ex·cre·tion (ek skrē′ shən) *noun* 1. A waste substance that is pushed out of an organism. 2. The act of pushing out such waste.

ex·cre·to·ry sys·tem (ek′ skri tôr ē sis′ təm) *noun* A group of organs that work together to remove waste from the body.

ex·hale (eks hāl′) *verb* To push air out of the body. See also *inhale.*

ex·o·skel·e·ton (ek sō skel′ i tən) *noun* A hard, jointed shell on the outside of an organism. See also *endoskeleton. Insects, crabs, and lobsters have exoskeletons.*

ex·o·ther·mic (ek sō thûr′ mik) *adjective* Describes a chemical reaction that gives off heat energy.

ex·pand (ek spand′) *verb* To spread apart into a larger space. See also *contract. Most materials expand when they are heated.*

ex·pel (ek spel′) *verb* To push out. *Jewelweeds expel their seeds when the seed pods burst open.*

ex·ten·sor (ek sten′ sər) *noun* A muscle that straightens or extends a body part. See also *flexor.*

ex·ter·nal fer·til·iz·a·tion (ek stûr′ nəl fûr tə lə zā′ shən) *noun* The result when egg and sperm unite outside the female's body. See also *internal fertilization.*

ex·tinct (ek stiŋkt′) *adjective* Describes a kind of organism with no living examples on Earth. **extinction** *noun*

ex·tra·ter·res·tri·al (ek strə te res′ trē əl) *adjective* Describes something that happens away from Earth or that comes from space.

Fahr·en·heit scale (far′ ən hīt skāl) *noun* Temperature scale commonly used in the United States, in which water freezes at 32° and boils at 212°.

fam·i·ly (fam′ ə lē) *noun* In classification of living things, a large group of similar animals, plants, or other organisms within an order.

fats (fats) *noun, plural* Oily food substances found in animals and some plants. *Fats give us warmth and energy, but too much fat in a person's diet can be harmful.*

fault (fôlt) *noun* A crack in Earth's crust along which earthquakes can happen. *Many earthquakes have occurred along the San Andreas fault in California.*

fault-block moun·tain (fôlt-block moun′ tən) *noun* A mountain formed when movement along cracks in Earth's crust pushes massive blocks of rock up or down.

fe·mur (fē′ mər) *noun* The thighbone.

fer·men·ta·tion (fûr men tā′ shən) *noun* The process by which organisms convert sugar into alcohol. *Fermentation turns apple juice into apple cider.* **ferment** *verb*

a	cat	e	net	îr	gear	u	cup	u̇	look, pull	*th*	this	ə	alive,
ā	day, lake	ē	seed	o	hot	ū	fuse	oi	soil		hw	wheel		comet,
ä	father	i	fit	ō	cold	ûr	fur, bird	ou	out		zh	measure		acid, atom,
âr	dare	ī	pine	ô	paw	ü	tool, rule	th	thin		ŋ	wing		focus

fern (fûrn) *noun*
A nonvascular plant with deeply lobed leaves, called fronds, that grows in shady, moist places. *Ferns are nonflowering plants that reproduce by spores.*

fer·til·i·za·tion (fûr te lə zā' shən) *noun*
1. The act of putting manure or other nutrient-rich material on plants to improve their health and increase their growth. 2. The uniting of a male and female sex cell of a plant or animal to begin a new individual. **fertilize** *verb*

fer·til·ized egg (fûr' tə līzd eg) *noun*
An egg, or female reproductive cell, that has joined with a male reproductive cell.

fer·til·iz·er (fûr' tə lī zər) *noun*
A substance that provides a plant with nutrients it needs to grow. *She used fertilizer to make her garden vegetables grow larger.*

fe·tus (fē' təs) *noun* A developing mammal that has its final form and structure but is not yet ready to be born.

fi·ber (fī' bər) *noun* 1. Threadlike materials that make up some tissues in plants or animals. 2. Indigestible material that helps keep food moving through the intestines.

fi·brous root (fī' brəs rüt) *noun* A root that branches in many directions. See also *taproot.*

fib·u·la (fib' yə lə) *noun* The outer and smaller of the two bones in the lower leg, between the knee and the ankle. See also *tibia.*

fil·a·ment (fil' ə mənt) *noun* An extremely thin wire or fiber. *An incandescent light bulb produces light with a glowing filament.*

fil·ter (fil' tər) 1. *noun* A material that cleans or strains substances that pass through it. 2. *verb* To clean or strain a gas or liquid by passing it through a filter.

fil·trate (fil' trāt) *noun* The liquid or gas that has passed through a filter.

fil·tra·tion (fil trā' shən) *noun* The process of passing through a filter.

fins (finz) *noun, plural* The parts of a fish's body that help it to swim, steer, and balance in the water.

First Law of Mo·tion (fûrst lô əv mō' shən) *noun* Isaac Newton's law that states that the velocity of an object can be changed only if an unbalanced force is applied to it.

first quar·ter Moon (fûrst kwôr' tər mün) *noun* The Moon phase when half of the lighted side is visible from Earth; sometimes called the half Moon. *The Moon cycle starts with the new Moon, and at the first quarter Moon phase, the Moon has made one-fourth of its full cycle.*

fish (fish) *noun* A cold-blooded vertebrate that has fins, gills, and scales. *Fish live in water and breathe through gills.*

fis·sion (fish' en) *noun* 1. The act of splitting the nucleus of an atom, which releases a huge amount of energy. 2. The dividing of a cell into two or more parts.

fixed pul·ley (fikst pül' ē) *noun* A simple machine consisting of a grooved wheel attached to a surface, so that the wheel spins but does not move. *They used a fixed pulley to lift the steel beam.*

fjord (fyôrd) *noun* A narrow inlet of the sea between cliffs or steep slopes, formed by glaciers.

flex·or (flek′ sər) *noun* A muscle that bends a body part. See also *extensor*.

flood plain (flud plān) *noun* The low-lying land near a stream or river that is covered by water during times of heavy rainfall or spring melting.

flow·er (flou′ ər) 1. *noun* In many plants, the structure that has the male and female sex cells and that often has showy blossoms. *Flowers can produce seeds and fruit.* 2. *verb* To produce flowers.

flow·er·ing plant (flou′ ər iŋ plant) *noun* A plant that produces flowers, which are its method of reproduction.

foam (fōm) *noun* 1. The frothy bubbles formed on a liquid, especially when the liquid is shaken or whipped at the surface. 2. A manufactured plastic or rubber material that has captured bubbles before becoming a solid.

fo·cal length (fō′ kəl leŋth) *noun* The distance from the surface of a curved mirror or the center of a lens to the point where the light rays come together.

fo·cal point (fō′ kəl point) *noun* The exact place where light rays come together after reflecting from a curved mirror or refracting through a lens.

fo·cus (fō kəs) 1. *noun* Another term for focal point. 2. *noun* The place within Earth's crust where an earthquake starts as rocks slide past one another. 3. *verb* To adjust a lens in order to make an image clear.

fog (fog) *noun* Tiny drops of water that make a thick whitish mist near the ground.

fold (fōld) 1. *noun* Something that is doubled over on itself. 2. *noun* The line or crease formed by something that is doubled over on itself. 3. *verb* To double something over on itself.

fold·ed moun·tain (fōld′ əd moun′ tən) *noun* A mountain formed when layers of rock are squeezed from the sides and pushed up into folds.

food chain (füd chān) *noun* The order in which animals feed on plants and on other animals. *Plants or algae form the base of almost all food chains.*

food cy·cle (füd sī′ kəl) *noun* The complete circle of events in which the matter in a living thing eventually is decomposed and returns to the soil, where it provides nutrients for plants and so continues in the cycle.

food group (füd grüp) *noun* Foods that contain the same types of nutrients. *Fruits and vegetables belong to the same food group.*

food guide pyr·a·mid (füd gīd pîr′ ə mid) *noun* A way of showing the food groups and the numbers of daily servings of each group recommended for a healthful diet.

a	cat	e	net	îr	gear	u	cup	u̇	look, pull	*th*	this	ə	alive,
ā	day, lake	ē	seed	o	hot	ū	fuse	oi	soil		hw	wheel		comet,
ä	father	i	fit	ō	cold	ûr	fur, bird	ou	out		zh	measure		acid, atom,
âr	dare	ī	pine	ô	paw	ü	tool, rule	th	thin		ŋ	wing		focus

food web (füd web) *noun* All of the connected and interacting food chains in an ecosystem.

foot-pound (füt′ pound) *noun* 1. The amount of work done by a force of 1 pound (16 ounces) moving over a distance of 1 foot (12 inches). 2. The amount of energy needed to move a 1-pound object a distance of 1 foot.

force (fôrs) *noun* Energy, such as a push or a pull, that changes the motion or shape of an object.

fore·cast (fôr′ kast) *noun* A prediction of the weather conditions for a given area at a given time.

fos·sil (fos′ əl) *noun* The remains or traces of an ancient living thing preserved in rock.

fossil fu·el (fos′ əl fül) *noun* A burnable source of energy that formed from prehistoric plants or animals. *Coal, oil, and natural gas are examples of fossil fuels.*

frac·ture (frak′ chər) *noun* A crack or a break in something, such as a rock or a bone.

free fall (frē fôl) *noun* The downward motion of a falling object caused by gravity and not slowed by other matter. *A skydiver experiences free fall until his parachute opens.*

freez·ing (frēz′ iŋ) *verb* To become solid as a result of decreasing temperature, as when water becomes ice.

freezing point (frēz′ iŋ point) *noun* The temperature at which a liquid becomes a solid. The freezing point of water is 0°C (32°F).

fre·quen·cy (frē′ kwən sē) *noun* The number of waves that pass by a point in a period of time, usually one second.

fric·tion (frik′ shən) *noun* The force, or resistance, caused when two things rub together.

frond (frond) *noun* A leaf that has many lobes or parts in it, usually on a fern or a palm.

front (frunt) *noun* In weather, the edge where two masses of air with different temperatures meet. *The approaching cold front resulted in a forecast of snow.*

frost (frost) *noun* The covering of tiny ice crystals that forms on surfaces when the temperature of moist air drops below the freezing point of water.

fruc·tose (fruk′ tōs) *noun* A form of sugar found in fruit juices and honey.

fruit (früt) *noun* 1. The part of a flowering plant that develops from the ovary and contains the seeds. 2. The edible part of a plant that is usually sweet and juicy and contains one or more seeds.

fu·el (fū′ əl) *noun* A substance that is burned or otherwise used as a source of heat or other energy.

ful·crum (fùl′ krəm) *noun* The point around which a lever turns or pivots. *A seesaw can be made from a board and a fulcrum.*

full Moon (fùl mün) *noun* The Moon phase when the whole lighted side is visible from Earth. *At the full Moon phase, the Moon has completed one-half of its 28-day cycle.*

fun·da·men·tal fre·quen·cy (fun də men′ təl frē′ kwən sē) *noun* The basic vibration of an object. *The fundamental frequency of a violin string is related to the musical note the string plays when it is not held down by a finger.*

fun·gi (fun' jī) *noun, plural* Multicelled, plantlike organisms that have no leaves, flowers or seeds, roots, or chlorophyll and that reproduce by spores. *Mushrooms, yeasts, and molds are examples of fungi.* **fungus** *singular*

fungi king·dom (fun' jī kiŋ' dəm) *noun* One of the five main classifications of living things, made up of fungi. *Members of the fungi kingdom are important decomposers.*

fuse (fūz) 1. *noun* A device in an electrical circuit that melts when too much current is flowing. *When a fuse melts, the circuit opens, electricity stops flowing, and a fire is prevented.* 2. *verb* To join or bond two materials by heating them to the melting point.

fu·se·lage (fū' sə läzh) *noun* The main or central body of an aircraft, which holds its passengers, crew, and cargo.

fu·sion (fū' zhən) *noun* The combining of the nuclei of two atoms to form another atom, which releases huge amounts of energy. *Stars are so bright because fusion occurs at very high pressures and temperatures inside them.*

fusion en·er·gy (fū' zhən en' ər jē) *noun* Energy released when two atoms of a lighter element, such as hydrogen, combine to form a heavier element, such as helium.

gal·ax·y (gal' ək sē) *noun* A massive collection of stars, dust, and gas held together by gravity. *The three types of galaxies are spiral, elliptical, and irregular.*

Ga·li·le·an sat·el·lites (gal ə lē' ən sat' ə līts) *noun* Jupiter's four largest moons (Ganymede, Europa, Callisto, and Io), first seen in 1610 by Galileo with a telescope.

gal·va·nom·e·ter (gal və nom' ə tər) *noun* An instrument that is able to detect or measure a small electric current.

gam·ete (gam' ēt) *noun* A mature male or female reproductive cell, able to reproduce if combined through fertilization.

gas (gas) *noun* A state of matter that does not have a definite shape or volume.

gas gi·ant (gas jī' ənt) *noun* A large planet made mostly of gases. *In our solar system, the gas giants are Jupiter, Saturn, Uranus, and Neptune.*

a	cat	e	net	îr	gear	u	cup	u̇	look, pull	*th*	**this**	ə	alive,
ā	day, lake	ē	seed	o	hot	ū	fuse	oi	soil	hw	**wheel**		comet,
ä	father	i	fit	ō	cold	ûr	**fur**, **bird**	ou	**out**	zh	measure		acid, atom,
âr	**dare**	ī	pine	ô	paw	ü	tool, rule	th	**thin**	ŋ	**wing**		focus

gears (gîrz) *noun, plural* A set of two or more wheels that have teeth, or cogs, that fit together so that when one wheel turns, the others also turn.

gel (jel) *noun* A jellylike, thick substance.

gem (jem) *noun* A cut and polished precious stone or a pearl, usually valued for its beauty.

gene (jēn) *noun* A specific section of a chromosome that controls a particular inherited trait or process. *Genes carry instructions for characteristics such as color of eyes that can be passed on to an individual's offspring.*

gen·er·a·tion (jen ə rā′ shən) *noun* 1. The average amount of time between when an organism is born until it is mature enough to reproduce. 2. A group of people who were born at about the same time. 3. The average span between parents and offspring. *Parents are one generation, and their offspring are another generation.*

gen·er·a·tor (jen′ ə rā tər) *noun* A machine that converts one form of energy into electrical energy. *Hydroelectric generators use the energy of falling water to make electric power.*

gene splic·ing (jēn splī′ siŋ) *noun* A process that joins or combines genetic material from different organisms or from different species.

ge·net·ic en·gi·neer·ing (jə net′ ik en jə nîr′ iŋ) *noun* The changing of genes in a laboratory to produce a particular trait in an organism.

ge·net·ics (jə net′ iks) *noun* The study of how the traits of an individual are passed on to its offspring through genes. **geneticist** *noun* **genetic** *adjective*

ge·nus (jē′ nəs) *noun* In classification of livings things, a grouping of closely related animals or plants within a family. *A genus is further divided into different species.*

ge·o·cen·tric mod·el (jē ō sen′ trik mod′ əl) *noun* An outdated model of the solar system that placed Earth at the center and the Sun and planets circling around it. *Copernicus was the first astronomer to suggest that the geocentric model was incorrect.*

ge·o·log·ic col·umn (jē ə loj′ ik kol′ əm) *noun* A series of Earth's rock layers in order from oldest to youngest.

ge·ol·o·gy (jē ol′ ə jē) *noun* The study of Earth, how it was formed, and the materials that make it up. **geologist** *noun*

ge·o·sta·tion·ar·y sat·el·lite (je ō stā′ shən er ē sat′ ə līt) *noun* An object that revolves around Earth at the same speed as Earth's rotation. *A geostationary satellite has a fixed orbit over the same place on Earth's surface.*

ge·o·ther·mal en·er·gy (jē ō thûr′ məl en′ ər jē) *noun* Heat energy that comes from Earth's interior and can be used to heat buildings or generate electricity.

germ (jûrm) *noun* 1. A microbe, such as a bacterium or virus, that can cause disease. 2. The seed, bud, or tiny beginning part from which an organism can develop.

ger·mi·nate (jûr mə nāt) *verb* To sprout and begin to grow into a new plant. *When a seed germinates, it grows shoots and roots.* **germination** *noun*

gills (gilz) *noun, plural* The organs that animals living in water use to get oxygen from the water.

ging·ko (giŋ′ kō) *noun* A large gymnosperm tree with fanlike leaves.

gla·cier (glā′ shər) *noun* A large body of year-round ice and snow that moves slowly across the land and deposits sediments where it melts.

gland (gland) *noun* A body part or organ that produces fluid that the body either uses (as with hormones) or discards (as with sweat).

glass·y (glas′ ē) *adjective* 1. Describes a solid substance that does not have a crystal structure. *Obsidian is a black, glassy rock formed when lava cools instantly as a volcano erupts.* 2. Describes a smooth, shiny surface luster of a mineral.

glid·er (glī′ dər) *noun* An aircraft that has no engine but flies in the air carried by wind currents.

glid·ing joint (glīd′ iŋ joint) *noun* A place where two bones meet that allows the bones to slide past each other but not to bend or rotate, for example, between the vertebrae.

glo·bal warm·ing (glō′ bəl wôrm′ iŋ) *noun* The slow increase of the average temperature of Earth's atmosphere.

glu·cose (glü′ kos) *noun* A natural sugar in plants.

gorge (gôrj) *noun* A narrow passage through high land, with steep, rocky slopes.

grad·u·at·ed cyl·in·der (graj′ ü āt əd sil′ ən dər) *noun* A tall, narrow container that has markings so that it can be used to measure the volume of liquids.

graft·ing (graf′ tiŋ) *noun* A form of asexual reproduction in which one plant part is attached to a different, rooted plant so they grow as one plant.

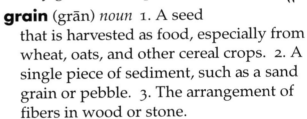

grain (grān) *noun* 1. A seed that is harvested as food, especially from wheat, oats, and other cereal crops. 2. A single piece of sediment, such as a sand grain or pebble. 3. The arrangement of fibers in wood or stone.

gram (gram) *noun* The basic unit of mass in the metric system. *One thousand grams equal one kilogram.*

grass·land (gras′ land) *noun* A biome whose main plants are grasses. *Large grazing animals such as bison are native to grasslands.*

grav·i·ta·tion·al po·ten·tial en·er·gy (grav i tā′ shən əl pə ten′ shəl en′ ər jē) *noun* The increase in potential energy caused when an object is lifted up against the force of gravity.

a	cat	e	net	îr	gear	u	cup	u̇	look, pull	*th*	this	ə	alive,
ā	day, lake	ē	seed	o	hot	ū	fuse	oi	soil	hw	wheel		comet,
ä	father	i	fit	ō	cold	ûr	fur, bird	ou	out	zh	measure		acid, atom,
âr	dare	ī	pine	ô	paw	ü	tool, rule	th	thin	ŋ	wing		focus

grav·i·trop·ism (grav i trōp' iz əm) *noun* A plant's response to gravity, usually by sending its roots downward.

grav·i·ty (grav' i tē) *noun* 1. The attraction that exists between any objects that have mass, pulling them together. 2. The force that pulls all objects toward the center of Earth. *When you jump up, it is gravity that pulls you back down.*

green·house ef·fect (grēn' hous ə fekt') *noun* The ability of certain gases in Earth's atmosphere, called greenhouse gases, to trap heat, keeping Earth warm.

greenhouse gas·es (grēn' hous gas' iz) *noun, plural* Carbon dioxide, methane, and similar polluting gases that can trap heat energy in Earth's atmosphere.

ground·wa·ter (ground' wôt ər) *noun* Water trapped underground in cracks and spaces in the bedrock that can come to the surface through wells or springs.

gym·no·sperm (jim' nə spûrm) *noun* A class of woody plants, such as conifers, that produce seeds that are not in a flower, pod, or fruit.

hab·it (hab' it) *noun* Something a person does regularly, often without thinking about it. *Brushing your teeth after meals is a healthy habit.*

hab·i·tat (hab' i tat) *noun* The place and surroundings where an animal or a plant normally lives.

hail (hāl) *noun* Precipitation that falls as small, round, layered pieces of ice that form inside a thunderstorm.

half-life (haf' līf) *noun* The amount of time it takes for half of a sample of radioactive material to decay to its next element. *The half-life of some elements can be hundreds of years.*

hap·loid cell (hap' loid sel) *noun* A cell that has half of a complete set of chromosomes, one chromosome from each pair. See also *diploid cell.*

hard·ness (härd' nəs) *noun* How resistant a mineral or other substance is to scratching.

haz·ard·ous (haz' ərd əs) *adjective* Dangerous or risky.

head·land (hed' lənd) *noun* A point of land that sticks out into a body of water.

health risk fac·tor (helth risk fak' tər) *noun* Something that makes it more likely than average that a person will have a certain medical problem. *Smoking is a health risk factor.*

heart (härt) *noun* The hollow, muscular organ that pumps blood through the body.

heat (hēt) *verb* To add heat energy to a substance.

heat en·er·gy (hēt en' ər jē) *noun* The energy a substance has because of the movement of the particles that it is made of.

heat ex·change (hēt eks chānj') *noun* The transfer of heat energy from one fluid to another without mixing or combining the fluids.

he·li·o·cen·tric mod·el (hē lē ō sen′ trik mod′ əl) *noun* The model of the solar system that places the Sun at the center. *Nicolaus Copernicus based his heliocentric model on observations made by another astronomer, Tycho Brahe.*

he·li·um (hē′ lē əm) *noun* A gas element that is less dense than air and will not burn. **Symbol: He**

he·lix (hē′ liks) *noun* A spiral or coil shape.

he·mo·di·al·y·sis (hē mō dī al′ ə sis) *noun* A method of cleaning the blood by removing it with medical equipment, filtering it, and then returning it to the body.

he·mo·glo·bin (hē′ mə glō bin) *noun* The part of red blood cells that contains iron and red pigment.

her·bi·vore (hûr′ bi vôr) *noun* An animal that eats only plants as food. **herbivorous** *adjective*

he·red·i·tar·y risk fac·tor (hə red′ i tər ē risk fak′ tər) *noun* A trait that has been inherited through the genes and that makes it more likely than average that a person will have a certain health problem.

he·red·i·ty (hə red′ i tē) *noun* The passing on of the parents' traits to their offspring. *Eye color and hair color are determined by heredity.* **hereditary** *adjective*

hertz (hûrtz) *noun* The unit of frequency for waves and vibrations. *One hertz equals one cycle per second.* **Abbreviation: Hz**

het·ero·ge·ne·ous (het ər ə jē′ nē əs) *adjective* Made up of unlike ingredients or parts. See also *homogeneous*.

hi·ber·na·tion (hī bər nā shən) *noun* A sleeplike state of lower body temperature and inactivity, which some animals use to survive winter.

high blood pres·sure (hī′ blud presh ər) *noun* Abnormally high readings of the pressure of blood against the walls of arteries; also called hypertension.

high·land (hī′ lənd) *noun* An area of mountains, hills, or plateaus.

high-pres·sure sys·tem (hī′ presh ər sis′ təm) *noun* A large, slowly swirling mass of sinking air that has high air pressure and fair weather.

hinge joint (hinj joint) *noun* A place where two bones meet that allows movement in only one direction, such as the knee.

hom·i·nid (hom′ ə nid) *noun* An organism in the family *Hominidae*, which are primates that walk upright on two feet.

ho·mo·ge·ne·ous (hō mə jē′ nē əs) *adjective* Made up of the same or similar parts. See also *heterogeneous*.

ho·ri·zon (hə rī′ zən) *noun* 1. The line where the sky appears to meet the land or the ocean. 2. One of a series of layers in soil.

a	cat	e	net	îr	gear	u	cup	u̇	look, pull	*th*	**this**	ə	alive,
ā	day, lake	ē	seed	o	hot	ū	fuse	oi	soil	hw	**wheel**		comet,
ä	father	i	fit	ō	cold	ûr	**fur**, bird	ou	**out**	zh	measure		acid, atom,
âr	dare	ī	pine	ô	paw	ü	**tool**, **rule**	th	**thin**	ŋ	wing		focus

hor·mone (hôr′ mōn) *noun* A chemical in the body made by a gland or organ that controls a certain body process, such as growth.

host (hōst) *noun* A living plant or animal on which a parasite lives and feeds. *Dogs and cats are often hosts for fleas.*

hot spot (hot spot) *noun* A place in the upper mantle of Earth's crust where magma wells up and forms a volcano.

hot spring (hot spriŋ) *noun* A place where water that is naturally heated by Earth's interior reaches the surface.

Hub·ble tel·e·scope (hub′ əl tel′ ə skōp) *noun* A telescope that is orbiting Earth and is used to observe distant objects in space.

hu·mid·i·ty (hū mid′ i tē) *noun* The amount of water vapor, or moisture, in the air. **humid** *adjective*

hu·mus (hū′ məs) *noun* The part of soil formed from rotting animal and vegetable matter that is dark and rich in nutrients.

hur·ri·cane (hûr′ i kān) *noun* A large, dangerous, low-pressure storm that forms over warm ocean water and has wind speeds of more than 118 kilometers (74 miles) per hour.

hy·brid (hī′ brid) *noun* The offspring of two related but different species of plants or animals. *A mule is a hybrid that is the offspring of a female horse and a male donkey.*

hy·drau·lic (hī drô′ lik) *adjective* Operated or moved by means of fluid forced through pipes by pressure.

hy·dro·car·bon (hī′ drō kär bən) *noun* An organic compound that contains only hydrogen and carbon.

hy·dro·e·lec·tric en·er·gy (hī drō i lek′ trik en′ ər jē) *noun* Electricity generated using the energy of water falling over a dam.

hydroelectric plant (hī drō i lek′ trik plant) *noun* The buildings and equipment used to produce hydroelectric energy.

hy·drom·e·ter (hī drom′ ə tər) *noun* An instrument used to measure a liquid's density.

hy·dro·pon·ics (hī drə pon′ iks) *noun* The practice of growing plants in nutrient-rich water instead of in soil.

hy·dro·sphere (hī′ drə sfîr) *noun* All the water on, under, and above Earth's surface.

hy·dro·trop·ism (hī drō trō′ piz əm) *noun* The tendency of a plant, especially its roots, to grow toward a source of water.

hy·per·son·ic (hī pər son′ ik) *adjective* Describes speed that is five or more times the speed of sound in air. See also *supersonic.*

hy·per·ther·mi·a (hī pər thûr′ mē ə) *noun* Extremely high fever, especially a fever caused artificially as part of medical treatment.

hy·poth·e·sis (hī poth′ ə sis) *noun* A possible explanation of why something happens. *Scientists conduct experiments to test a hypothesis.*

Ice Age (īs āj) *noun* A period of time in geological history when much of Earth's land surface was covered by glaciers. *The most recent Ice Age ended about 10,000 years ago.*

ice·berg (īs′ bûrg) *noun* A huge mass of ice that has broken off from a glacier and is floating in the ocean.

ig·ne·ous rock (ig′ nē əs rok) *noun* A type of rock formed from molten rock (magma) that cooled.

im·age (im′ ij) *noun* A picture of an object formed when light rays that bounced off that object pass through a lens or reflect off a mirror. *Lenses in microscopes produce images that are larger than the original objects.*

im·i·ta·tion (im i tā′ shən) *noun* A likeness or copy of something.

im·mune (i mūn′) *adjective* Protected against a disease, either by vaccination or by having survived the disease and thus having resistance to it. **immunity** *noun*

immune sys·tem (i mūn′ sis′ təm) *noun* The body system that protects against disease and infection. *Antibodies and white blood cells are important parts of the body's immune system.*

im·per·fect flow·er (im pûr′ fekt flou′ ər) *noun* A flower that has only stamens or a pistil but not both. See also *perfect flower.*

imprint 1. (im′ print) *noun* A kind of fossil made when the impression of an object in sediment is preserved in sedimentary rock. 2. (im print′) *verb* To fix something firmly in memory.

in·clined plane (in′ klīnd plān) *noun* A simple machine made of a flat, slanted surface, or ramp.

in·com·plete dom·i·nance (in kəm plēt′ dom′ i nəns) *noun* The genetic pattern in which both forms of a trait come through rather than one trait masking, or dominating, the other.

incomplete flow·er (in kəm plēt′ flou′ ər) *noun* A flower that normally is missing either sepals, petals, stamens, or pistils. See also *complete flower.*

in·dex fos·sil (in′ deks fos′ əl) *noun* A fossil of an organism that lived in such a narrow time range that it is used to identify related geological formations.

a	cat	e	net	îr	gear	u	cup	u̇	look, pull	*th*	**this**	ə	alive,
ā	day, lake	ē	seed	o	hot	ū	fuse	oi	soil	hw	**wheel**		comet,
ä	father	i	fit	ō	cold	ûr	fur, bird	ou	**out**	zh	measure		acid, atom,
âr	dare	ī	pine	ô	paw	ü	tool, rule	th	**thin**	ŋ	wing		focus

in·di·cat·or (in′ di kā tər) *noun* 1. A substance that changes color in the presence or absence of another substance. *Red litmus paper is an indicator because it turns blue when it touches a base.* 2. An organism that is present only under certain conditions and so can be used to show or find those conditions. *Organisms that live only in clean water are indicators of unpolluted water.*

in·di·vid·u·al (in di vij′ ü əl) *noun* A single organism, such as an individual human. *A population is made up of the total number of individuals in a certain area.*

in·er·tia (i nûr′ shə) *noun* The tendency of an object to resist a change in its movement, whether the object is moving or at rest. *A ball's inertia will cause it to stay in one place unless you give it a push to start it rolling.*

in·fec·tion (in fek′ shən) *noun* A disease caused by bacteria or viruses.
infectious *adjective*

in·fec·tious di·sease (in fek′ shəs di zēz′) *noun* An illness that can spread from one person to another.

in·flam·ma·tion (in flə mā′ shən) *noun* The presence of swelling, pain, redness, and a feeling of higher temperature caused by an injury or a disease.

in·for·ma·tion su·per·high·way (in fər mā′ shən sü pər hī′ wā) *noun* The Internet, a vast electronic communication network linking facts, data, and people that is available to any computer user who has the equipment to log onto it.

in·fra·red ra·di·a·tion (in′ frə red rā dē ā′ shən) *noun* A wavelength of electromagnetic energy with a frequency just below the visible spectrum of light (just below the color red).

in·got (iŋ′ gət) *noun* A block or bar of metal, shaped that way to make it easy to store or transport.

in·hale (in hāl′) *verb* To breathe in, or draw air into the lungs. See also *exhale.*

in·her·it·ed trait (in her′ i təd trāt) *noun* A specific characteristic or trait that was received through heredity from an individual's parents.

in·ner core (in′ ər kôr) *noun* The innermost, solid layer of Earth.

inner ear (in′ ər îr) *noun* The innermost part of the ear, where vibrations are sent along nerves to the brain.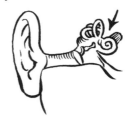

inner plan·et (in′ ər plan′ it) *noun* Any of the four planets in our solar system closest to the Sun: Mercury, Venus, Earth, or Mars.

in·sol·u·ble (in sôl′ yə bəl) *adjective* Describes a substance that will not dissolve in a particular liquid. See also *soluble.*

in·stinct (in′ stiŋkt) *noun* A way of behaving that is natural, not learned.

in·su·late (in′ sə lāt) *verb* To cover something with a material in order to block or reduce the transfer of heat energy, electricity, or sound. *Electrical wires are often insulated with plastic.*

in·su·la·tion (in sə lā' shən) *noun*
A material that is used to block or reduce the transfer of heat, electricity, or sound.

in·su·lat·or (in sə lā' tər) *noun* A material that blocks or reduces the flow of heat, electricity, or sound. See also *conductor*.

in·ten·si·ty (in ten' si tē) *noun*
The strength of a force or energy.

in·ter·act (in tər akt') *verb* To act on one another. *The members of a pack of wolves will interact with each other.*

in·ter·nal com·bus·tion en·gine
(in tûr' nəl kəm bus' chən en' jin) *noun*
A machine or motor that burns a fuel and transfers the energy of the fuel into the energy of motion. *Most automobiles use internal combustion engines.*

internal fer·til·i·za·tion (in tûr' nəl fûr tə lə zā' shən) *noun* Union of the male and female reproductive cells that occurs within the female. See also *external fertilization*.

In·ter·na·tion·al Date Line
(in tûr nash' ə nəl dāt līn) *noun*
The imaginary line (at 180° longitude) that runs approximately from the North to the South Poles, directly opposite the prime meridian. *The International Date Line shows the place where each calendar day begins.*

In·ter·net (in' tûr net) *noun* The electronic communications network that makes worldwide data and information available to anybody with a computer equipped to use it.

in·ter·tid·al zone (in tûr tī' dəl zōn) *noun* The coastal area between the lowest low-tide mark and the highest high-tide mark, which is a habitat for many plants and animals.

in·verse re·la·tion·ship (in vûrs' ri lā' shən ship) *noun* A direct opposite, for example, a connection between variables in which one increases as the other decreases.

in·ver·te·brate (in vûr' tə brāt) *noun* An animal that does not have a backbone. See also *vertebrate*.

in·vol·un·tar·y mus·cle (in vôl' ən ter ē mus' əl) *noun* A body tissue that automatically contracts to produce motion, without the organism thinking about it. See also *voluntary muscle*.

i·on (ī' ən, ī' on) *noun* An atom that is charged because it has either gained or lost an electron.

i·ris (ī' ris) *noun*
The colored ring around the pupil that expands or contracts to control the amount of light entering the eye.

ir·reg·u·lar gal·ax·y (i reg' yə lər gal' ək sē) *noun* A massive collection of stars, dust, and gas that has an overall shape that is not symmetrical.

is·land arc (ī' lənd ärk) *noun* A group of volcanic islands made when molten rock rises from beneath the sea floor. *The Hawaiian Islands are an island arc that formed in the Pacific Ocean.*

a	cat	e	net	îr	gear	u	cup	ù	look, pull	*th*	**this**	ə	alive,
ā	day, lake	ē	seed	o	hot	ū	fuse	oi	soil		hw	**wheel**		comet,
ä	father	i	fit	ō	cold	ûr	fur, bird	ou	**out**	zh	measure		acid, atom,
âr	dare	ī	pine	ô	paw	ü	tool, rule	th	**thin**	ŋ	wing		focus

i·so·bar (ī′ sə bär) *noun* A line on a weather map connecting areas that have the same air pressure at the same time.

i·so·la·tion (ī sə lā′ shən) *noun* The state of being alone, not with other organisms.

i·so·mers (ī′ sə mərs) *noun, plural* Two or more carbon-hydrogen compounds that have the same chemical formula but whose molecules have different shapes.

i·so·topes (ī′ sə tōps) *noun, plural* Two or more atoms of the same element that contain the same number of protons but have different numbers of neutrons in their nuclei.

jet en·gine (jet en′ jin) *noun* An engine, especially an aircraft engine, whose power comes from pushing a jet of heated gases out behind it.

jet·ty (jet′ ē) *noun* A seawall or pier built out into the water.

joint (joint) *noun* A place where two bones join together or meet. *There are four types of joints in your body: immovable, gliding, hinge, and ball-and-socket.*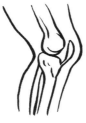

joule (jül) *noun* The unit of energy and work in the metric system. *One joule equals the work done by a force of 1 newton over a distance of 1 meter.*

Ju·pi·ter (jü′ pi tər) *noun* The fifth planet from the Sun and the largest planet in the solar system.

kelp (kelp) *noun* A large, coarse seaweed that is used as food, fertilizer, and a source of iodine.

Kel·vin scale (kel′ vin skāl) *noun* The temperature scale used in physics and chemistry that starts at absolute zero, where particles have no motion: 0 Kelvins equals −273.15°C or −459.4°F.

Kep·ler's laws (kep′ lərz lôz) *noun, plural* Three scientific laws that describe the motion of planets, formulated by Johannes Kepler.

kid·ney (kid′ nē) *noun* One of the pair of organs in the body that filter waste matter from the blood, turning it into urine.

ki·lo·me·ter (ki lom′ ə tər, kil′ ə mē tər) *noun* A metric unit of length, equal to 1,000 meters. **Abbreviation: km**

ki·lo·me·ters per hour (ki lom′ ə tərz, kil′ ə mē tərz pər our) *noun* A measurement of speed that describes how many kilometers would be traveled in one hour at the rate the object is moving. **Abbreviation: km/h**

kil·o·watt (kil′ ə wot) *noun* A unit of electric power equal to 1,000 watts.

kilowatt-hour (kil' ə wot our) *noun*
A unit of work or energy, mainly used to describe electrical energy used, equal to 1,000 watts of power in one hour.

ki·net·ic en·er·gy (kə net' ik en' ər jē) *noun* Energy that an object has because it is moving.

king·dom (kiŋ' dəm) *noun* One of the five major classifications of living things. *The five kingdoms are animals, plants, fungi, monerans, and protists.*

Kui·per Belt (kī' pər belt) *noun* An area in our solar system reaching to about 15 billion kilometers (9.3 billion miles) beyond Pluto's orbit and likely containing hundreds of thousands of cometlike objects.

lac·tose (lak' tōs) *noun* A form of sugar that is found in milk.

lake (lāk) *noun* A large body of fresh water that does not flow and is surrounded by land.

land breeze (land brēz) *noun* Air that moves from a landmass toward a body of water at night.

land·form (land' fôrm) *noun* A natural feature of Earth's surface. *Mountains and valleys are examples of landforms.*

La Ni·ña (lə nēn' yə) *noun* An unusually cold ocean current in the eastern Pacific Ocean near the equator. See also *El Niño.*

large in·tes·tine (lärj in tes' tin) *noun* The lower, wider part of the digestive system, below the small intestine, where water is removed from solid waste.

lar·va (lär' və) *noun* The young, wormlike life cycle stage of some insects, between egg and pupa.

lar·ynx (lar' iŋks) *noun* The upper end of the trachea (windpipe), where the vocal cords are located.

la·ser (lā' zər) *noun* A device that sends out a light beam in which all the waves have the same wavelength and travel parallel to each other rather than spreading out. *Lasers are used in some kinds of surgery and also in light shows.*

last quar·ter Moon (last kwôr' tər mün) *noun* The waning phase of the Moon when about half of the lighted side is visible from Earth; sometimes called a half Moon.

lat·i·tude (lat' i tüd) *noun* The distance north or south of the equator, measured in degrees.

a	cat	e	net	îr	gear	u	cup	u̇	look, pull	*th*	**th**is	ə	alive,
ā	day, lake	ē	seed	o	hot	ū	fuse	oi	soil	hw	**wh**eel		comet,
ä	father	i	fit	ō	cold	ûr	fur, bird	ou	out	zh	mea**s**ure		acid, atom,
âr	dare	ī	pine	ô	paw	ü	tool, rule	*th*	**th**in	ŋ	wi**ng**		focus

la·va (lä′ və) *noun* The extremely hot, melted rock that pours out of an erupting volcano. *When magma reaches Earth's surface, it is called lava.*

law of con·ser·va·tion of en·er·gy (lô əv kon sər vā′ shən əv en′ ər jē) *noun* The scientific law stating that the total amount of energy in a system does not change but transfers from one form to another.

law of conservation of mass (lô əv kon sər vā′ shən əv mas) *noun* The scientific law stating that mass cannot be created or destroyed in a chemical reaction; also called *the law of conservation of matter.*

law of conservation of mo·men·tum (lô əv kon sər vā′ shən əv mə men′ təm) *noun* The scientific law stating that when two objects collide, the total momentum does not change but is only transferred. See also *momentum.*

law of re·flec·tion (lô əv ri flek′ shən) *noun* The scientific law stating that a wave bounces off a surface at the same angle at which it hits the surface but in the opposite direction.

law of u·ni·ver·sal grav·i·ta·tion (lô əv ū nə vûr′ səl grav i tā′ shən) *noun* The scientific law stating that a force of attraction exists between all objects in the universe that is related to the mass of the objects and the distance between them.

Laws of Mo·tion (lôz əv mō′ shən) *noun, plural* Three scientific laws that describe the motion of objects, formulated by Isaac Newton.

leaf (lēf) *noun* A flat, usually green part of a plant that grows from the stem. *The leaf contains chlorophyll and is where photosynthesis takes place.* **leaves** *plural*

learned be·hav·ior (lûrnd bi hāv′ yər) *noun* Behavior that an animal has been taught or has learned from experience. See also *instinct.*

leg·end (lej′ ənd) *noun* The notes on a map or chart that give information such as the scale used and the meanings of symbols. *They used the legend on the weather map to find the symbol for precipitation.*

lens (lenz) *noun* The clear, curved part of the eye inside the iris that focuses light on the retina.

lev·er (lev′ ər) *noun* A simple machine made of a bar that turns on a fixed point, or fulcrum.

life cy·cle (līf sī′ kəl) *noun* The series of stages and changes that living things experience from birth to death. *The life cycle of many insects includes the egg, larva, pupa, and adult stages.*

life span (līf span) *noun* The number of years an organism is expected to live, based on the average number of years that others of its species live.

lift (lift) *noun* The upward force on the wing of a bird or an aircraft as it is moving. *The shape of an airplane wing produces lift by reducing the air pressure above it.* **lift** *verb*

lig·a·ment (lig′ ə mənt) *noun* Tissue that connects bone to bone at a joint and that holds some organs in place.

light (līt) *noun* Energy in the form of waves that the human eye sees as visible light.

light·ning (līt′ niŋ) *noun* A discharge of electric current in the sky that passes from cloud to cloud or from a cloud to Earth, producing a bright flash of light.

light ray (līt rā) *noun* A straight line that shows the direction light waves are traveling.

light-year (līt'-yîr) *noun* A unit of distance equal to the distance light travels in space in one year, or about 9.5 trillion kilometers (about 6 trillion miles).

lig·nite (lig' nīt) *noun* A form of coal in which the texture of the original plants from which it formed is visible.

lim·it·ing fac·tor (lim' it iŋ fak' tər) *noun* Something in an environment that keeps the population of an organism from increasing as much as it could. *Air temperature is a limiting factor for different kinds of plants.*

lip·id (lip' id) *noun* One of the three main building blocks of cells.

liq·uid (lik' wid) *noun* A state of matter that has a definite volume but no definite shape. *A liquid takes the shape of its container.*

li·ter (lē' tər) *noun* The basic metric unit of volume or capacity, equal to 1,000 milliliters. **Abbreviation: L**

lith·o·sphere (lith' ə sfîr) *noun* The outer, solid part of Earth including the crust plus the upper part of the mantle.

liv·ing (liv' iŋ) *adjective* Not dead; having life, or alive.

load (lōd) *noun* The weight or other resistance force of the object being moved using a simple machine. *You push down on one end of a lever to lift a load at the other end.*

lo·cal winds (lō' kəl windz) *noun, plural* Winds that form in a small area, for example, a wind coming off the water in the daytime.

lo·co·mo·tion (lō kə mō' shən) *noun* The ability to move from place to place. *Locomotion is a basic need of most animals.*

lode·stone (lōd' stōn) *noun* A mineral that contains iron and acts as a magnet.

long-day plant (loŋ' dā plant) *noun* A plant that flowers or grows in response to a long period of daylight.

lon·gi·tude (lon' ji tüd) *noun* The distance east or west of the prime meridian, an imaginary line that runs from the North to South Pole through Greenwich, England, measured in degrees.

low-pres·sure sys·tem (lō' presh ər sis' təm) *noun* A large, swirling mass of rising air that has low air pressure and that usually brings wet, stormy weather.

lu·bri·ca·tion (lü brə kā' shən) *noun* The use of a substance such as grease or oil to reduce friction and make machine parts glide and run more smoothly. **lubricant** *noun*

lu·nar (lü' nər) *adjective* Relating to the Moon.

lunar e·clipse (lü' nər ē klips') *noun* A temporary blocking of the Moon that happens when Earth passes between the Sun and the Moon, putting the Moon in Earth's shadow.

a	cat	e	net	îr	gear	u	cup	u̇	look, pull	*th*	this	ə	alive,
ā	day, lake	ē	seed	o	hot	ū	fuse	oi	soil	hw	wheel		comet,
ä	father	i	fit	ō	cold	ûr	fur, bird	ou	out	zh	measure		acid, atom,
âr	dare	ī	pine	ô	paw	ü	tool, rule	th	thin	ŋ	wing		focus

lung (luŋ) *noun* One of the main organs of the respiratory system, where oxygen enters the bloodstream and waste gases are removed.

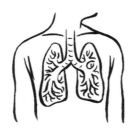

lung vol·ume (luŋ vol′ ūm) *noun* The capacity of the lungs, or the amount of air that a person's lungs can hold.

lus·ter (lus′ tər) *noun* The brightness or shine of a mineral's surface. *Glassy and metallic are two examples of luster.*

Mm

ma·chine (mə shēn′) *noun* A device that uses energy and does work. See also *simple machine, complex machine.*

mac·ro·scop·ic (mak rə skop′ ik) *adjective* Large enough to be seen by the naked eye.

mag·ma (mag′ mə) *noun* The molten rock beneath Earth's surface. *When magma flows out of a volcano, it is called lava.*

mag·net (mag′ nit) *noun* Any material that has a magnetic field around it and so attracts iron and steel.

mag·net·ic field (mag net′ ik fēld) *noun* The space around a magnet where magnetic force attracts iron or steel.

magnetic re·ver·sal (mag net′ ik ri vûr′ səl) *noun* The switching of Earth's north and south magnetic poles. *Magnetic reversal has taken place many times throughout geological history.*

mag·net·ism (mag′ ni tiz əm) *noun* The force of a magnet to pull iron or steel toward itself.

mag·ni·fy (mag′ ni fī) *verb* To make larger. **magnification** *noun*

mag·ni·tude (mag′ ni tüd) *noun* 1. The brightness of a star. See also *absolute magnitude, apparent magnitude.* 2. The amount of energy released during an earthquake. See also *Richter scale.*

main se·quence (mān sē′ kwens) *noun* A group of stars on a diagram, used to demonstrate the many stages a star passes through over its lifetime.

main-sequence star (mān sē′ kwens stär) *noun* A star that fuses hydrogen into helium.

mal·le·a·ble (mal′ ē ə bəl) *adjective* Describes a metal that is easily bent or shaped.

mam·mal (mam′ əl) *noun* A warm-blooded vertebrate whose females produce milk to feed their young.

man·di·ble (man′ də bəl) *noun* The lower jaw bone.

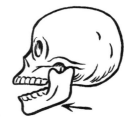

man·tle (man′ təl) *noun* The partly molten layer inside Earth between the crust and the outer core.

map (map) *noun* A drawing that shows features of a surface, especially Earth's surface.

ma·re (mä′ rā) *noun, singular* A large, dark, mostly flat area on the surface of the Moon, formed by ancient lava flows. **maria** *plural*

mar·row (mâr′ ō) *noun* A soft, red or yellow, fatty substance that fills the hollow centers of bones and makes blood cells.

Mars (märz) *noun* The fourth planet from the Sun, a reddish-colored, terrestrial planet.

mass (mas) *noun* The amount of matter in a substance, measured in grams. *An object's mass does not change.*

mass wast·ing (mas wāst′ iŋ) *noun* The downhill movement of Earth materials, such as soil and rocks, that is caused by gravity; also called *mass movement.*

mat·ter (mat′ ər) *noun* Anything that occupies space and has mass. *Solids, liquids, and gases are all forms of matter.*

ma·ture (mə chür′) *adjective* Fully developed or fully grown and able to reproduce. *A female kitten becomes a mature cat at about 7 months of age.*

meas·ure (mezh′ ər) *verb* To find the area, mass, size, temperature, volume, or weight of an object or the time of an event.

me·chan·i·cal ad·van·tage (mə kan′ i kəl ad van′ tij) *noun* The number of times a machine multiplies the effort force, often expressed as a ratio.

mechanical en·er·gy (mə kan′ i kəl en′ ər jē) *noun* Energy related to the motion or potential motion of objects.

mechanical weath·er·ing (mə kan′ i kəl we*th*′ ər iŋ) *noun* The wearing away of a rock or other material by the physical actions of ice, water, wind, and plants. See also *chemical weathering.*

me·di·an (mē′ dē ən) *noun* In a set of values listed in order, the value that is in the middle, or halfway point. *In the number set 2, 2, 4, 5, 7, 8, 12, the median is 5.*

me·di·um (mē′ dē əm) *noun* A substance through which energy moves or is carried.

me·dul·la (mə dul′ ə) *noun* 1. The deep, inner part of a plant or animal structure. 2. Short for medulla oblongata, an inner part of the brain.

mei·o·sis (mī ō′ sis) *noun* The process by which the number of chromosomes in a reproductive cell divides in half. *Meiosis prepares a reproductive cell for combining with a reproductive cell of the opposite sex.* See also *mitosis.*

mel·a·nin (mel′ ə nin) *noun* The dark brown or black pigment that is responsible for skin, eye, and hair color.

melt (melt) *verb* To change from a solid to a liquid as a result of increasing temperature, for example, when ice becomes water.

melt·ing point (melt′ iŋ point) *noun* The temperature at which a solid becomes liquid, which for ice is 0°C (32°F).

a	cat	e	net	îr	gear	u	cup	u̇	look, pull	*th*	this	ə	alive,
ā	day, lake	ē	seed	o	hot	ū	fuse	oi	soil	hw	wheel		comet,
ä	father	i	fit	ō	cold	ûr	fur, bird	ou	out	zh	measure		acid, atom,
âr	dare	ī	pine	ô	paw	ü	tool, rule	th	thin	ŋ	wing		focus

mem·brane (mem′ brān) *noun* A very thin layer of tissue that covers a cell, an organ, or another body part.

men·o·pause (men′ ə poz) *noun* The time in a female's life when she reaches an age at which she is no longer able to bear children.

Mer·cu·ry (mûr′ kyə rē) *noun* The planet closest to the Sun and the second smallest planet in our solar system.

Mes·o·zo·ic er·a (mez ə zō′ ik îr′ ə) *noun* The period of geological history between 245 and 65 million years ago, when dinosaurs lived and birds and mammals first appeared on Earth.

me·tab·o·lism (mə tab′ ə liz əm) *noun* The process by which plants and animals release energy from food and get rid of waste products.

met·al (met′ əl) *noun* A chemical element that is usually hard and shiny and can be melted, hammered, and shaped. *Metals are usually good conductors of electricity and heat energy.*

met·a·mor·phic rock (met ə môr′ fik rok) *noun* A rock that has been changed by heat or pressure into another kind of rock.

met·a·mor·pho·sis (met ə môr′ fə sis) *noun* A change in form or structure that takes place in the life cycle of certain animals. *A tadpole goes through metamorphosis to become an adult frog.*

me·te·or (mē′ tē ər) *noun* A piece of rock or metal falling through a planet's atmosphere and burning brightly because of friction.

me·te·or·ite (mē′ tē ə rīt) *noun* A part of a meteor that falls to the surface of a planet or moon without completely burning up.

me·te·or·oid (mē′ tē ə roid) *noun* A piece of rock or metal, smaller than an asteroid, that orbits the Sun or a planet. *When a meteoroid hits a planet's atmosphere, it burns up as a meteor.*

me·te·o·rol·o·gy (mē tē ə rol′ ə jē) *noun* The study of Earth's atmosphere, including its weather and climate.

me·ter (mē′ tər) *noun* The basic unit of length or distance in the metric system, equal to 100 centimeters, or about 39.37 inches.

met·ric (met′ rik) *adjective* Related to the metric system of measurement, which is based on units of 10.

metric rul·er (met′ rik rü′ lər) *noun* A measuring device that is a thin, straight piece of wood, plastic, or metal marked with centimeters and millimeters.

metric sys·tem (met′ rik sis′ təm) *noun* A system of measurement based on units of 10.

mi·cro·cli·mate (mī′ krō klī mət) *noun* Conditions of temperature, light, and humidity that exist within a small area or habitat. *The shady side of a building can be a microclimate in which mosses can grow.*

mi·cro·or·gan·ism (mī krō ôr′ gə nizm) *noun* A living thing, such as a bacterium, so small that it can only be seen with a microscope; also called a microbe.

mi·cro·scope (mī′ krə skōp) *noun* An optical instrument that uses a combination of lenses for viewing objects that are too small to see with the naked eye.

mi·cro·scop·ic (mī krə skop′ ik) *adjective* Describes something so small that it can only be seen through a microscope.

mid·dle ear (mid' əl îr) *noun* The section of the ear behind the eardrum containing several small bones (the hammer, the anvil, and the stirrup) that transmit sound vibrations to the inner ear.

mid·o·cean ridge (mid' ō shən rij) *noun* An underwater mountain chain that circles Earth, extending through the middle of most oceans. *New ocean crust is created at midocean ridges.*

mi·gra·tion (mī grā' shən) *noun* The regular movement of some animals from one region to another for feeding or breeding.

miles per hour (mīlz pər our) *noun* A measurement of speed that describes how many miles would be traveled in one hour at the rate the object is moving. **Abbreviation: mph**

Milk·y Way (mil' kē wā) *noun* The name of the galaxy to which our solar system belongs. *The Milky Way contains more than 100 billion stars and can be seen on a dark night as a broad band of faint light across the sky.*

mil·li·li·ter (mil' ə lē tər) *noun* The metric unit of volume or capacity equal to one-thousandth of a liter. *Five milliliters equals about one teaspoon.* **Abbreviation: mL**

mim·ic·ry (mim' i krē) *noun* An organism's resemblance to another organism or to an object in its environment that helps protect it from danger.

min·er·al (min' ər əl) *noun* A solid, naturally occurring substance that was not made by plants or animals. *Minerals can combine to form rocks.*

mi·to·chon·dri·a (mī tō kon' drē ə) *noun, plural* Rod-shaped structures found in almost all cells that supply energy to the cell. *Mitochondria release energy through cell respiration.* **mitochondrion** singular

mi·to·sis (mī tō' sis) *noun* The formation of two new cell nuclei in a dividing cell, each of which has the same number of chromosomes as the original nucleus. See also *meiosis.*

mix·ture (miks' chər) *noun* A combination of two or more different substances that keep their own properties and can be separated again.

mod·el (mod' əl) *noun* In science, a way of showing something in order to better understand how it works. *We use a model of the solar system to study planetary orbits.*

mold (mōld) *noun* 1. A fossil made when a hollow shape left by an organism fills with minerals and hardens. *When the rock specimen was split, a mold of a shell was visible.* See also *cast.* 2. A type of tiny fungus that grows in moist places.

mol·e·cule (mol' ə kūl) *noun* Two or more atoms chemically bonded together.

mol·lusk (mol' əsk) *noun* A soft-bodied animal without a backbone that usually lives in water and has a shell. *Oysters, clams, and nautiluses are mollusks.*

a	cat	e	net	îr	gear	u	cup	u̇	look, pull	*th*	this	ə	alive,
ā	day, lake	ē	seed	o	hot	ū	fuse	oi	soil	hw	wheel		comet,
ä	father	i	fit	ō	cold	ûr	fur, bird	ou	out	zh	measure		acid, atom,
âr	dare	ī	pine	ô	paw	ü	tool, rule	th	thin	ŋ	wing		focus

molt (mōlt) *verb* To shed an outer covering of skin, feathers, hair, or shell before a new covering grows. *Lobsters molt as they grow larger.*

mo·men·tum (mō men′ təm) *noun* The amount of force or energy an object has when it is moving.

mon·er·an (mon′ ə ran) *noun* A one-celled organism that lacks a distinct cell nucleus. *Bacteria are monerans.*

moneran king·dom (mon′ ə ran kiŋ′ dəm) *noun* The scientific grouping of one-celled organisms that have no distinct nucleus. *The moneran kingdom is one of the five main classifications of living things.*

mon·o·cot (mon′ ə kot) *noun* Short for monocotyledon, a plant that has one seed leaf. See also *dicot, seed leaf. Grasses and corn are examples of monocots.*

moon (mün) *noun* A natural, rocky object that revolves around a planet.

mo·raine (mə rān′) *noun* A mound or ridge of unsorted rock debris deposited at the edges of a glacier as it melts.

Morse code (môrs kōd) *noun* A system of communication that uses dots and dashes transmitted by either light or sound.

moss (mos) *noun* A kind of nonvascular plant that grows in moist, shady places. *There are over 35,000 species of moss.*

mo·tion (mō′ shən) *noun* A change in position compared to the starting location, which happens over time.

mo·tor (mō′ tər) *noun* A device that transfers electrical energy into mechanical energy. *The electric pencil sharpener uses a motor to spin the cutting wheels.*

mo·tor neu·ron (mō′ tər nùr′ on) *noun* A nerve cell that controls movement, or motor activity, in the body.

moun·tain (moun′ tən) *noun* A part of Earth's surface that is much higher than the surrounding land. *Mount McKinley is the highest mountain in the United States.*

mov·a·ble pul·ley (mü′ və bəl pùl′ ē) *noun* A pulley attached to the load that is being lifted and that moves freely along the rope. *The workers stood on the roof and used a movable pulley to lift the piano.*

mu·cus (mū′ kəs) *noun* The slimy, thick fluid that coats and protects the inside of various body parts, especially the mouth, nose, and throat.

mul·ti·cel·lu·lar (mul tē sel′ yə lər) *adjective* Made up of many cells.

mus·cle (mus′ əl) *noun* A body tissue made up of bundles of fibers that stretch and contract to move parts of the body.

mus·cu·lar sys·tem (mus′ kyə lər sis′ təm) *noun* The body system that works to cause movement.

mu·ta·tion (mū tā′ shən) *noun* A change in genes that produces a trait in an animal or plant that did not appear in its parents.

mu·tu·al·ism (mū′ chü ə liz əm) *noun* A relationship between two species that is helpful to both species.

nar·cot·ic (när kot′ ik) *noun* A pain-relieving drug that can be addictive when misused.

nat·u·ral gas (nach′ ər əl gas) *noun* A mixture of gases found beneath Earth's surface and used as fuel for cooking and heating.

natural re·source (nach′ ər əl rē′ sôrs) *noun* A material found in nature that is valuable to humans, such as water, wood, and coal.

natural se·lec·tion (nach′ ər əl sə lek′ shən) *noun* The very slow, ongoing natural process in which the best suited organisms survive to reproduce and carry on their species. See also *evolution*.

na·ture (nā′ chər) *noun* Everything natural in the universe. *All living things, landforms, and weather make up what we call nature.*

neap tide (nēp tīd) *noun* The time of smallest change between low tide and high tide. See also *spring tide*. *Neap tides occur when the Sun, Moon, and Earth line up to form a right angle.*

near-shore zone (nîr′ shôr zōn) *noun* The area of ocean water that extends outward from shore for a short distance.

neb·u·la (neb′ yə lə) *noun* A gigantic cloud of dust and gas that exists in the space between stars and is the first stage of star formation. **nebulae** *plural*

nek·ton (nek′ tən) *noun, plural* Marine organisms that are free-swimming, moving in aquatic ecosystems in the middle depths of the sea, where wave patterns and currents do not affect them.

neph·ron (nef′ ron) *noun* In the kidneys, a single excretory cell or unit.

Nep·tune (nep′ tün) *noun* The eighth planet from the Sun in our solar system.

nerve (nûrv) *noun* A body tissue through which electrical impulses pass.

nerve im·pulse (nûrv im′ puls) *noun* A message carried by a nerve.

ner·vous sys·tem (nûr′ vəs sis′ təm) *noun* The bodily system that, in vertebrates, controls most of the body's actions and functions. *The nervous system includes the brain, spinal cord, and nerves.*

net force (net′ fôrs) *noun* The combined effect of all the forces that act on an object.

neu·ron (nur′ on) *noun* The basic unit of nervous tissue; also called a nerve cell.

neu·tral·i·za·tion (nü trə lī zā′ shən) *noun* The process of chemically adjusting the pH of a substance so that it is neither acid nor base.

a	cat	e	net	îr	gear	u	cup	u̇	look, pull	th	this	ə	alive,
ā	day, lake	ē	seed	o	hot	ū	fuse	oi	soil	hw	wheel		comet,
ä	father	i	fit	ō	cold	ûr	fur, bird	ou	out	zh	measure		acid, atom,
âr	dare	ī	pine	ô	paw	ü	tool, rule	th	thin	ŋ	wing		focus

neu·tron (nü′ tron) *noun* A microscopic particle that is located in the nucleus of an atom and has no electric charge.

neutron star (nü′ tron stär) *noun* A very dense, tiny star that is nearing the end of its life cycle.

new Moon (nü mün) *noun* The first phase of the Moon, when it is difficult to see it in the sky because its lighted side is facing the Sun.

new·ton (nü′ tən) *noun* The metric unit used to measure force.

NEX·RAD (neks′ rad) *noun* A form of Doppler radar used for tracking storms. Stands for NEXt generation of RADar.

niche (nich) *noun* 1. A habitat that supplies everything needed for a species or an organism to survive. 2. The role played by a living thing or species in a food web.

nic·o·tine (nik′ ə tēn) *noun* The addictive, poisonous substance found in tobacco.

ni·tro·gen (nī′ trə jən) *noun* The odorless, colorless gas that makes up about 78 percent of Earth's atmosphere by volume. **Symbol: N**

nitrogen cy·cle (nī′ trə jən sī′ kəl) *noun* The continuous series of processes in which nitrogen passes from the air into the soil, then into an organism, and then back into the air.

no·ble gas (nō′ bəl gas) *noun* Any of a group of extremely stable gases, such as helium.

noise pol·lu·tion (noiz pə lü′ shən) *noun* Extreme sound that is annoying or even harmful in an environment.

non·com·mu·ni·ca·ble dis·ease (non kə mū′ ni kə bəl di zēz′) *noun* An illness that cannot be passed from one person to another, such as heart disease.

non·liv·ing (non liv′ iŋ) *adjective* Not alive and never having been alive. *Minerals are nonliving objects because they have never been alive.*

non·re·new·a·ble re·source (non rē nü′ ə bəl rē′ sôrs) *noun* Something of value, use, or need that cannot be replaced or renewed by natural cycles or by good management.

non·vas·cu·lar (non vas′ kyə lər) *adjective* Lacking the vessels that carry blood, sap, or other fluids in an organism.

North·ern Hem·i·sphere (nôr*th* ərn hem′ i sfîr) *noun* The half of Earth that is north of the equator.

North Pole (nôrth pōl) *noun* The northernmost point on Earth, located at the northern tip of Earth's axis.

no·va (nō′ və) *noun* An exploding star that suddenly gets much brighter and then gradually fades to its original level of magnitude.

nu·cle·ar en·er·gy (nü′ klē ər en′ ər jē) *noun* Force from a radioactive source or from a nuclear reaction.

nuclear fis·sion (nü′ klē ər fish′ ən) *noun* The splitting of an atomic nucleus into two smaller nuclei, which releases huge amounts of energy that can be used to produce electricity.

nuclear fu·sion (nü′ klē ər fū′ zhən) *noun* The joining of atomic nuclei to form heavier ones, which releases energy that may someday be used to produce electricity.

nuclear mem·brane (nü′ klē ər mem′ brān) *noun* The thin layer of tissue that encloses a cell nucleus.

nu·cle·ic ac·id (nü klē′ ik as′ id) *noun* Any of the acids that are made up of nucleotide chains, such as DNA and RNA, that enable cells to build proteins.

nu·cle·us (nü′ klē əs) *noun* 1. The center, or core, of an atom, where most of its mass is. 2. The main part of a cell, which controls the cell's activities.

nu·mer·a·tor (nü′ mə rā tər) *noun* The part of a fraction that is above the line. See also *denominator*.

nu·tri·ent (nü′ trē ənt) *noun* A substance that is a needed source of nourishment or nutrition for a plant or an animal. *Vitamins, proteins, and minerals are examples of nutrients.*

nu·tri·tion (nü trish′ ən) *noun* 1. The complete process by which a plant or animal takes in and uses food for its survival, health, and growth. 2. The science or study of properly balanced diets to promote health, especially in humans. *Good nutrition involves eating a variety of healthy foods.*

nymph (nimf) *noun* The young form of some insects that looks much like a small adult. *The dragonfly nymph lives in water and does not have wings.*

o·bese (ō bēs′) *adjective* Extremely overweight.

ob·serv·a·to·ry (əb zûrv′ ə tôr ē) *noun* A building that is equipped for viewing and studying the sky, stars, or weather.

oc·clud·ed front (ə klüd′ əd frunt) *noun* The boundary line where a cold front catches up with a warm front.

o·cean (ō′ shən) *noun* 1. All of the salt water that covers more than two-thirds of Earth's surface. 2. One of the four main parts into which the world ocean is divided: Arctic, Atlantic, Indian, and Pacific Oceans.

o·ce·an·ic crust (ō shē an′ ik krust) *noun* The part of Earth's crust that makes up the ocean floor.

o·cean·og·ra·phy (ō shə nog′ rə fē) *noun* The study of the oceans, including their depth and extent, water chemistry, marine biology, and resource use.
oceanographer *noun*

oc·tave (ok′ təv) *noun* 1. The musical interval between eight tones in the scale. 2. The whole series of eight tones that make up the modern musical scale.

a	cat	e	net	îr	gear	u	cup	u̇	look, pull	*th*	**this**	ə	alive,
ā	day, lake	ē	seed	o	hot	ū	fuse	oi	soil	hw	**wheel**		comet,
ä	father	i	fit	ō	cold	ûr	fur, bird	ou	out	zh	measure		acid, atom,
âr	dare	ī	pine	ô	paw	ü	tool, rule	th	thin	ŋ	wing		focus

ohm (ōm) *noun* A unit used to measure the resistance of a substance to the flow of electricity through it.

oil (oil) *noun* A thick, greasy, liquid substance that burns easily and is used as a fuel, lubricant, or food. *Oil comes from many sources, including animals, plants, and fossils.*

ol·fac·to·ry nerve (ôl fak' tə rē nûrv) *noun* The nerve that carries information about smells from the nose to the brain.

om·ni·vore (om' nə vôr) *noun* A consumer that eats both plants and animals as food.

one-celled or·gan·ism (wun' seld ôr' gə niz əm) *noun* A living thing, such as an amoeba, that is made up of only a single cell.

o·paque (ō pāk') *adjective* Not letting any light through.

o·pen cir·cuit (ō' pən sûr' kit) *noun* A circuit in which electric current cannot flow because the path is not complete, or not closed. See also *closed circuit.*

open o·cean (ō' pən ō' shən) *noun* A vast expanse of the sea, away from land and deeper than 200 meters (650 feet).

open-ocean zone (ō' pən ō' shən zōn) *noun* The part of the sea that includes most deep ocean waters, where most organisms live near the surface.

op·tic nerve (op' tik nûrv) *noun* The nerve that carries information about what we see from the eye to the brain.

or·bit (ôr' bit) *noun* The path that an object in space follows around another object in space, for example, Earth's orbit around the Sun. **orbit** *verb*

or·der (ôr' dər) *noun* 1. In classification, a group of closely related plants or animals within a class. 2. A way things are placed or organized in a study or an experiment.

ore (ôr) *noun* A mineral or rock material that is mined to get a substance that it contains, such as gold, silver, or iron.

or·gan (ôr' gən) *noun* A specialized part or structure of an organism, such as a leaf or stem of a plant, or the heart or brain of an animal.

or·gan·elle (ôr gə nel') *noun* A specialized part of a cell that functions in a way an organ would function in an animal.

or·gan·ism (ôr' gə niz əm) *noun* Any living thing, including all plants and animals.

or·gan sys·tem (ôr' gən sis' təm) *noun* A set of organs that work together in a living thing to do a certain job.

o·rig·i·nal hor·i·zon·tal·i·ty (ə rij' ə nəl hôr ə zən tal' ə tē) *noun* The theory that many rocks form flat, horizontal layers.

os·mo·sis (oz mō' sis) *noun* The process by which a fluid passes through a membrane, for example, water through a cell membrane.

out·crop (out' krop) *noun* A mass of bedrock exposed at Earth's surface.

out·er core (out' ər kôr) *noun* The outermost part of Earth's core, a liquid layer between the mantle and the solid inner core.

out·er ear (out' ər îr) *noun* The part of the ear that sticks out from the head, used to pick up sound vibrations and direct them into the middle ear; also called the ear flap.

out·er plan·et (out′ ər plan′ it) *noun* Any of the five planets in our solar system farthest from the Sun: Jupiter, Saturn, Uranus, Neptune, or Pluto.

out·put force (out′ pu̇t fôrs) *noun* The amount of energy actually delivered by a machine or system in response to the input force applied.

o·va·ry (ō′ və rē) *noun* The female organ that produces eggs in animals and seeds in plants.

o·ver·pop·u·la·tion (ō vər pop yə lā′ shən) *noun* A situation in which too many animals compete for the natural resources in an area.

o·ver·tone (ō′ vər tōn) *noun* One of a series of higher pitches that sounds along with the fundamental note. *Overtones combine with the basic pitch to give sound its quality.*

o·vi·duct (ō′ və dukt) *noun* A tube through which eggs travel from an ovary to other parts of the reproductive system.

o·vule (ō′ vūl) *noun* In a flowering plant, the part of the ovary that develops into a seed.

ox·i·da·tion (ok si dā′ shən) *noun* A chemical reaction that occurs when a substance is exposed to oxygen. *Oxidation causes iron to rust.*

ox·y·gen (ok′ sə jen) *noun* The colorless, odorless, tasteless gas that makes up almost 20 percent of Earth's air and is necessary for life. **Symbol: O**

oxygen–car·bon di·ox·ide cy·cle (ok′ sə jen kär′ bən dī ok′ sīd sī′ kəl) *noun* The ongoing process by which oxygen in the atmosphere is changed into carbon dioxide through animal breathing, and carbon dioxide is changed into oxygen by green plants through photosynthesis.

o·zone (ō′ zōn) *noun* A form of oxygen that is pale blue and has a sharp smell. *Ozone is a major pollutant in the lower atmosphere, but it is helpful in the upper atmosphere.* **Formula: O_3**

ozone lay·er (ō′ zōn lā′ ər) *noun* The layer of mostly ozone gas in the upper atmosphere that blocks much of the Sun's harmful ultraviolet light from reaching Earth.

pa·le·on·tol·o·gist (pā lē ən tol′ ə jist) *noun* A scientist who studies the remains of living things from the ancient past. **paleontology** *noun*

Pa·le·o·zo·ic er·a (pā lē ə zō′ ik îr′ ə) *noun* The period of geological history from about 545 million to about 245 million years ago, when the first fish, insects, and amphibians appeared on Earth.

a	cat	e	net	îr	gear	u	cup	u̇	look, pull	*th*	**this**	ə	alive,
ā	day, lake	ē	seed	o	hot	ū	fuse	oi	soil		hw **wheel**		comet,
ä	father	i	fit	ō	cold	ûr	fur, bird	ou	out		zh measure		acid, atom,
âr	dare	ī	pine	ô	paw	u̇	tool, rule	th	thin		ŋ wing		focus

pal·i·sade lay·er (pal ə sād′ lā′ ər) *noun* A layer of chloroplast cells under the surface of a leaf, where photosynthesis takes place.

Pan·gae·a (pan jē′ ə) *noun* The huge landmass that existed about 225 million years ago and that very slowly split apart to form all of today's continents.

par·al·lax (par′ ə laks) *noun* The change in the view of an object when seen from two different points—for example, when looking with just the right eye and then just the left eye. *Astronomers use parallax to estimate distances to stars, viewing them from different points in Earth's orbit.*

par·al·lel cir·cuit (par′ ə ləl sûr′ kit) *noun* An electrical circuit in which each device is connected separately to the cell so the current travels in several paths.

par·a·site (par′ ə sīt) *noun* An organism that lives in or on another organism, called the host. See also *host*.

par·a·sit·ism (par′ ə si tiz əm) *noun* The relationship in which one organism lives in or on, and benefits from, another organism, which may be harmed by the relationship.

pas·sive trans·port (pas′ iv trans pôrt) *noun* The movement of molecules through a cell membrane without the use of energy. See also *active transport*.

pa·tel·la (pə tel′ ə) *noun* The triangular bone in front of the knee joint that forms the kneecap.

path·o·gen (path′ ə jen) *noun* An organism that causes disease.

peat (pēt) *noun* Partly decayed plant matter usually found in bogs or swamps.

ped·i·gree (ped′ ə grē) *noun* A chart that traces the history of traits in a family.

pel·vis (pel′ vəs) *noun* The basin-shaped framework of bones that supports the lower abdomen, formed by the hipbones and lower backbone. **pelvic** *adjective*

pen·i·cil·lin (pen ə sil′ ən) *noun* An antibiotic that fights disease, first developed in 1928 by Sir Alexander Fleming from a type of mold.

pe·ren·ni·al (pə ren′ ē əl) *adjective* Describes a plant that lives for several years, with new growth occurring each year. **perennial** *noun*

per·fect flow·er (pûr′ fikt flou′ ər) *noun* A flower that has both male and female parts (stamens and pistil). See also *imperfect flower.*

per·i·gee (per′ ə jē) *noun* The point in the orbit of a satellite or planet when it is nearest to the planet or star it is circling.

pe·ri·od (pîr′ ē əd) *noun* 1. A division of geological time, larger than an epoch but smaller than an era. *The Mesozoic era includes the Jurassic and Triassic periods.* 2. A term used to describe the similar chemical properties shared by certain elements.

pe·ri·od·ic ta·ble (pîr ē od′ ik tā′ bəl) *noun* The chart of elements arranged in increasing order of atomic number and grouped by similar properties.

per·ish (per′ ish) *verb* To die, or to become destroyed or ruined. *Some scientists believe a meteorite hitting Earth caused all the dinosaurs to perish.*

per·ma·frost (pûr′ mə frost) *noun* Ground that is always frozen.

per·me·a·bil·i·ty (pûr mē ə bil′ i tē) *noun* The rate at which water can pass through a material. **permeable** *adjective*

per·spi·ra·tion (pûr spə rā′ shən) *noun*
1. Water and salts secreted, or given off, by glands in the skin. 2. The process of giving off these secretions. **perspire** *verb*

pes·ti·cide (pes′ ti sīd) *noun* A chemical that kills insects or other animals believed to be harmful.

pet·al (pet′ əl) *noun* The colored or showy part of a flower.

pet·ri·fi·ca·tion (pet′ rə fi kā′ shən) *noun* The natural preservation of plant or animal parts by being turned to stone.

pet·ri·fied fos·sil (pet′ rə fīd fos′ əl) *noun* Plant or animal remains whose cells have been replaced by minerals.

pH (pē āch) *noun* A scale for indicating the strength of an acid, on which neutral water has a pH of 7, acids have a pH below 7, and bases have a pH above 7. Stands for potential of Hydrogen.

phase of the Moon (fāz əv *th*ə mün) *noun* A stage in the apparent changing shapes of the Moon as it revolves around Earth.

phlo·em (flō′ əm) *noun* The tissue in vascular plants that carries food from the leaves to the rest of the plant. See also *xylem. Phloem usually carries material downward in a plant.*

pho·no·graph (fōn′ ə graf) *noun* An instrument for reproducing sounds recorded on the grooves of a record.

phos·phate (fos′ fāt) *noun* A compound containing phosphorus and oxygen, widely used in fertilizers.

pho·ton (fō′ ton) *noun* A particle of electromagnetic radiation. *Light energy is made up of streams of photons.*

pho·to·pe·ri·od·ism (fō tō pîr′ ē ə diz əm) *noun* An organism's response to changes in the length of daylight and darkness.

pho·to·sphere (fō′ tə sfîr) *noun* The visible surface of the Sun or other star.

pho·to·syn·the·sis (fō tō sin′ thə sis) *noun* The process by which green plant cells make food from sunlight, water, and carbon dioxide.

pho·to·tro·pism (fō tə trō′ piz əm) *noun* The response of a plant to changes in light. **phototropic** *adjective*

phy·lum (fī′ ləm) *noun* In the classification of living things, a very large group of plants or animals, such as mollusks or arthropods. **phyla** *plural*

phys·i·cal change (fiz′ i kəl chānj) *noun* A change in the properties of a substance that does not create a new kind of matter. See also *chemical change.*

physical fit·ness (fiz′ i kəl fit′ nəs) *noun* Good health that is a result of eating a nutritious diet, exercising, and having other healthy habits.

physical prop·er·ty (fiz′ i kəl prop′ ər tē) *noun* A characteristic of matter, such as size, shape, or state, that can be observed and measured.

a	cat	e	net	îr	gear	u	cup	u̇	look, pull	*th*	**this**	ə	alive,
ā	day, lake	ē	seed	o	hot	ū	fuse	oi	soil	hw	**wheel**		comet,
ä	father	i	fit	ō	cold	ûr	fur, bird	ou	**out**	zh	measure		acid, atom,
âr	dare	ī	pine	ô	paw	ü	tool, rule	th	**thin**	ŋ	wing		focus

phy·to·plank·ton (fī tō plaŋk′ tən) *noun*
Tiny floating plant life in a body of water.
See also *zooplankton*.

pig·ment (pig′ mənt) *noun* The substance
in cells and tissues that gives color to
plants and animals.

pi·o·neer com·mu·ni·ty (pī ə nîr′
ke mū′ ni tē) *noun* The first community to
thrive in a once-lifeless habitat.

pioneer plant (pī ə nîr′ plant) *noun*
The first plant species to grow in a
bare area.

pioneer spe·cies (pī ə nîr′ spē′ shēz)
noun The first kind of organism to live
and grow in an area.

pis·til (pis′ təl) *noun*
The female part of a
flowering plant,
located in the center of
a flower and including
the stigma, style, and
ovary.

pitch (pich) *noun* How high or low a sound
is, determined by the frequency of sound
waves. *A dog whistle has a pitch too high for
the human ear to hear.*

plain (plān) *noun* A large, flat area of land
with few trees.

plane mir·ror (plān mîr′ ər) *noun*
A mirror with a reflecting surface that is
flat, not concave or convex.

plan·et (plan′ it) *noun* A large celestial
body that revolves around a star.
planetary *adjective*

plank·ton (plaŋk′ tən) *noun, plural* Tiny
living things that float or drift in ocean
water and are food for many sea animals.

plant be·hav·ior (plant bi hāv′ yər) *noun*
The response of plants to changing
conditions or to stimuli.

plant king·dom (plant kiŋ′ dəm) *noun*
One of the five main classifications of
living things, made up of multicelled
organisms with cell walls that need
sunlight, water, and minerals, and do not
move or have sensory systems.

plas·ma (plaz′ mə) *noun* The clear liquid
part of blood.

plate (plāt) *noun* One of the huge, slowly
moving blocks of rock that make up
Earth's crust.

pla·teau (pla tō′) *noun* A large area of flat
land higher than the surrounding land.

plate bound·ar·y (plāt boun′ də rē)
noun The line or edge where two plates
meet, often the location of earthquakes,
volcanoes, and mountain formation.

plate·let (plāt′ lət) *noun* A tiny body in
blood that helps it clot to stop bleeding.

plate tec·ton·ics (plāt tek ton′ iks) *noun*
The widely accepted theory that today's
continents were once part of a single
landmass that broke apart about 200
million years ago, have moved into their
present locations, and are still in motion.
See also *Pangaea*.

Plu·to (plü′ tō) *noun* The farthest planet
from the Sun and the smallest planet in
our solar system. *Pluto appears to be a
small, rocky planet covered with ice.*

po·lar·i·ty (pō lâr′ ə tē) *noun*
The characteristic of having opposite
properties at opposite sides or ends.
Polarity is a physical property of magnets.

po·lar·i·za·tion (pō lər ə zā′ shən) *noun*
Allowing light vibrations to pass through
in only one direction.

pole (pōl) *noun* 1. Either end of the imaginary axis that runs through Earth. *The poles are the northernmost and southernmost points on Earth.* 2. One of the two ends or sides of a magnet where the magnetic force is strongest. *The north pole of one magnet is attracted to the south pole of another magnet.* **polar** *adjective*

pol·len (pol' ən) *noun* Dustlike grains on the stamen of a flower that contain the male reproductive cells.

pol·li·na·tion

(pol ə nā' shən) *noun* The transfer of pollen from the male stamen to the female stigma of a flowering plant as part of reproduction.

pol·li·na·tor (pol' ə nā tər) *noun* An animal, such as a bee or butterfly, that carries pollen from one plant to another so that pollination can occur.

pol·lut·ant (pə lüt' ənt) *noun* A substance that spoils, dirties, or damages some aspect of the environment.

pol·lute (pə lüt) *noun* To add unnatural substances to the air, land, or water.

pol·lu·tion (pə lü' shən) *noun* The result of adding harmful pollutants, such as wastes, smoke, gases, chemicals, and pesticides, to the environment. *Pollution can make air unhealthy to breathe and water unsafe to drink.*

pol·y·mer (pol' ə mər) *noun* A chemical compound made by linking smaller molecules in a long, repeating chain.

pop·u·la·tion (pop yə lā' shən) *noun* All of the members of one species that live in one area. **populate** *verb*

pore (pôr) *noun* 1. A tiny opening in the skin that allows sweat and oils to come out. 2. A tiny space between grains in sand or soil. **porous** *adjective* **porosity** *noun*

po·si·tion (pə zish' ən) *noun* An object's place or location.

po·ten·tial en·er·gy (pə ten' shəl en' ər jē) *noun* 1. The energy available as a result of an object's position or condition. *You have potential energy as you stand at the edge of a swimming pool ready to jump in.* 2. Any kind of stored energy.

pow·er (pou' ər) *noun* 1. The level of magnification of a microscope. 2. The energy or force to do work.

prai·rie (prâr' ē) *noun* A large area of flat or hilly land covered with grasses and flowering plants.

Pre·cam·bri·an er·a (prē kam' brē ən îr' ə) *noun* The earliest era of Earth's geological history, when there was no life, before the Paleozoic era.

pre·cip·i·ta·tion (pri sip i tā' shən) *noun* Water that falls to Earth, usually as rain, snow, sleet, or hail, as part of the water cycle.

pred·a·tor (pred' ə tər) *noun* An animal that hunts, kills, and eats other animals.

pre·dict (pri dikt') *verb* To state in advance what you think the outcome of an event or experiment will be. *Meteorologists predict tomorrow's weather.*

a	cat	e	net	îr	gear	u	cup	ů	look, pull	*th*	**this**	ə	alive,
ā	day, lake	ē	seed	o	hot	ū	fuse	oi	soil		hw **wheel**		comet,
ä	father	i	fit	ō	cold	ûr	**fur**, bird	ou	**out**		zh measure		acid, atom,
âr	dare	ī	pine	ô	paw	ü	**tool**, rule	*th*	**thin**		ŋ wing		focus

pre·dic·tion (pri dik′ shən) *noun* A statement of what you think will happen in the future.

pre·scribe (pri skrīb′) *verb* To write an order for the preparation and use of a medicine. **prescription** *noun*

pres·er·va·tion (prez ər vā′ shən) *noun* Efforts to protect an area, such as a wilderness area, so that it stays in its original state.

pres·sure (presh′ ər) *noun* The force produced by pushing or pressing on something, such as air pressure or blood pressure.

pre·vail·ing wind (pri vāl′ iŋ wind) *noun* A global wind that usually blows in the same direction.

prey (prā) *noun* An animal that is hunted, killed, and eaten by a predator animal.

pri·ma·ry col·or (prī′ mer ē kul′ ər) *noun* A color that cannot be made by mixing other colors. *The primary colors of light beams are red, blue, and green.*

primary con·su·mer (prī′ mer ē kən sü′ mər) *noun* An animal that feeds on producers, that is, green plants. See also *secondary consumer.*

primary pig·ment (prī′ mer ē pig′ mənt) *noun* A coloring substance that cannot be made by mixing other colors. *The primary pigments in paints and inks are red, blue, and yellow.*

primary wave (prī′ mer ē wāv) *noun* A back-and-forth wave of energy released in an earthquake that causes back-and-forth vibrations of rocks in the same direction the wave moves. **Abbreviation: P wave**

prime me·rid·i·an (prīm mə rid′ ē ən) *noun* The imaginary north-south line marking 0° longitude that passes through Greenwich, England.

prism (priz′ əm) *noun* A piece of clear glass or plastic, with parallel bases and three or more sides, that bends or breaks up light into the colors of the spectrum.

prob·a·bil·i·ty (prob ə bil′ i tē) *noun* A measure of how likely something is to happen, usually expressed as a percentage or ratio. *The probability that a newborn will be a male is about 1 in 2, or 50 percent.*

pro·duc·er (prə dü′ sər) *noun* An organism that makes its own food as well as food for consumers. *Green plants are the producers in most food chains.* **produce** *verb*

prod·uct (prod′ ukt) *noun* A new substance that is the result of a chemical change.

prop·er·ty (prop′ ər tē) *noun* A characteristic of matter that can be observed and measured, such as volume, density, or mass.

prop root (prop rüt) *noun* A root that supports or holds up the main plant.

pro·tec·tive col·or·a·tion (prə tek′ tiv kul ə rā′ shən) *noun* A kind of camouflage in which the color of an organism helps it blend in with its background. *An Arctic hare's white coat is an example of protective coloration.*

pro·tein (prō′ tēn) *noun* One of many of the carbon compounds present in all living cells and needed for cell growth and repair.

pro·tist (prō′ tist) *noun* A kind of one-celled organism with a nucleus.

protist king·dom (prō′ tist kiŋ′ dəm) *noun* One of the five main classifications of living things, made up of protists such as algae, protozoa, and diatoms.

pro·ton (prō′ ton) *noun* A particle in the nucleus of an atom that carries a positive electric charge.

pro·to·star (prō′ tə stär) *noun* A young star that glows as gravity causes it to collapse.

pro·to·zo·an (prō tə zō′ ən) *noun* A tiny one-celled organism, such as an amoeba, that reproduces by dividing.

pu·ber·ty (pū′ bər tē) *noun* The time of rapid physical growth in adolescent humans when boys and girls change from children to adults.

pul·ley (půl′ ē) *noun* A simple machine made of a wheel with a groove in the rim through which a chain or rope runs.

pul·sar (pul′ sär) *noun* A rapidly spinning star that sends out bursts of radio waves, which make it seem to be "blinking." Stands for PULSating radio stAR.

pulse (puls) *noun* The regular beating of the arteries as the heart pumps blood through them. **pulse** *verb*

Pun·nett square (pun′ it skwâr) *noun* A diagram used to calculate the possible combinations of traits in offspring of the same parents.

pu·pa (pū′ pə) *noun* The stage in the life cycle of most insects between larva and adult, in which metamorphosis occurs while the pupa is inside a cocoon.

pu·pil (pū′ pəl) *noun* The round opening in the iris through which light enters the eye. *Your pupil looks like a black dot in the center of your eye.*

pure·bred (pūr′ bred) *adjective* Describes a self-pollinating organism that passes on the same traits to all its offspring for several generations.

pyr·a·mid (pîr′ ə mid) *noun* A solid shape with a square base and triangular sides that meet at a point at the top.

quad·rant (kwod′ rənt) *noun* A quarter of a circle or a quarter of the circumference of a circle.

qual·i·ty (kwol′ i tē) *noun* The difference you hear between two sounds of the same volume and pitch.

quar·ry (kwôr′ ē) *noun* 1. A deep pit where stone is dug or blasted out of the earth. 2. A hunted animal, or prey.

a	cat	e	net	îr	gear	u	cup	ů	look, pull	*th*	**this**	ə	alive,
ā	day, lake	ē	seed	o	hot	ū	fuse	oi	soil	hw	**wheel**		comet,
ä	father	i	fit	ō	cold	ûr	fur, bird	ou	**out**	zh	**measure**		acid, atom,
âr	dare	ī	pine	ô	paw	ü	tool, rule	th	**thin**	ŋ	**wing**		focus

qua·sar (kwā′ zär) *noun* An extremely bright starlike object very far from Earth that gives off huge amounts of electromagnetic radiation. Stands for QUASi-stellAR.

queen (kwēn) *noun* The egg-laying female in a colony of insects, such as ants or bees.

ra·dar (rā′ där) *noun* An electronic device for tracking the position and path of a distant moving object. Stands for RAdio Detecting And Ranging. *Radar works by sending out radio waves and recording the echoes reflected back by objects.*

rad·i·al sym·me·try (rā′ dē əl sim′ ə trē) *noun* A balanced arrangement of parts around a center point.

ra·di·ant en·er·gy (rā′ dē ənt en′ ər jē) *noun* The kind of energy carried by electromagnetic waves.

ra·di·a·tion (rā dē ā′ shən) *noun* The transfer of thermal energy by electromagnetic waves.

ra·di·a·tive balance (rā′ dē ə tiv bal′ əns) *noun* A balance between energy lost and energy gained.

ra·di·o·ac·tive el·e·ment (rā dē ō ak′ tiv el′ ə mənt) *noun* An element such as uranium or plutonium whose atomic nuclei naturally give off high-energy particles and rays.

ra·di·o·te·le·scope (rā dē ō tel′ ə skōp) *noun* A telescope that uses several large dish antennas to pick up natural radio waves from space.

ra·di·um (rā′ dē əm) *noun* An element that is a silvery white, highly radioactive metal. **Symbol: Ra**

rain·bow (rān′ bō) *noun* An arc of colored bands across the sky after a rainstorm caused by water droplets in the air bending sunlight into its separate colors.

rain for·est (rān fôr′ əst) *noun* A tropical region of very tall trees, much rainfall, and great diversity of life. *The plants of the rain forests produce much of Earth's oxygen.*

rain gauge (rān gāj) *noun* An instrument used to measure rainfall.

rar·e·fac·tion (râr ə fak′ shən) *noun* The part of a sound wave where the molecules are spread apart. See also *compression.*

ra·ti·o (rā′ shē ō) *noun* A mathematical comparison of two quantities or numbers.

raw ma·te·ri·al (rô mə tîr′ ē əl) *noun* A substance that must be treated or processed to be made into a useful finished product. *Crude oil is the raw material from which we get petroleum.*

ray (rā) *noun* A narrow beam of radiation, such as light or heat rays.

re·ac·tant (rē ak′ tənt) *noun* The original substance that is acted on and changed in a chemical reaction.

re·ac·tion force (rē ak' shən fôrs) *noun*
The force that pushes or pulls back against an action force, in Newton's Third Law of Motion. See also *action force*.

reaction time (rē ak' shən tīm) *noun* The time between a stimulus and a response.

re·ac·tiv·it·y (rē ak tiv' i tē) *noun* The ability of a substance to go through a chemical change.

rea·son·ing (rē' zə niŋ) *noun* The process of thinking in a systematic way and drawing conclusions based on facts.

re·cep·tor (ri sep' tər) *noun* A sense organ or sensory cell that reacts to a stimulus and sends a signal to a nerve.

re·cess·ive fac·tor (ri ses' iv fak' tər) *noun* The trait masked or hidden in offspring when the factors in a pair are different.

recessive gene (ri ses' iv jēn) *noun* In a pair of genes, the one that is masked if the dominant gene is present. See also *dominant gene*.

recessive trait (ri ses' iv trāt) *noun* A trait that is not expressed or shown in the offspring. See also *dominant trait*.

rec·la·ma·tion (rek lə mā' shən) *noun* The process of restoring a damaged ecosystem, for example, by planting trees after a forest fire.

re·con·struc·tion (rē kən struk' shən) *noun* The process of rebuilding a broken or destroyed object.

re·cy·cle (rē sī' kəl) *verb* To reuse old objects or process them so they can be made into new products. *Plastics, paper, and aluminum cans are often recycled.*

red blood cell (red blud sel) *noun* A cell in the blood that carries oxygen to the body cells and carries away waste gases.

red gi·ant (red jī' ənt) *noun* A very old star that is cooler, larger, less dense, and brighter than the Sun.

re·duce (ri düs') *verb* To lessen, or make smaller.

re·flect (ri flekt') *verb* To bounce off a surface. **reflection** *noun*

re·flect·ing tel·e·scope (ri flekt' iŋ tel' ə skōp) *noun* A telescope that uses a concave mirror to collect and focus light from objects in space.

re·flex (rē' fleks) *noun* An automatic reaction to a stimulus that happens without thought or effort.

re·fract (ri frakt') *verb* To bend light rays as they pass from one substance into another, such as from air into water. **refraction** *noun*

re·fract·ing tel·e·scope (re frak' tiŋ tel' ə skōp) *noun* A telescope that uses two convex lenses to collect and focus light from objects in space.

re·gen·er·ate (rē jen' ər āt) *verb* To grow a new body part to replace a damaged or lost one. *Starfish can regenerate new arms.*

re·gen·er·a·tion (rē jen ə rā' shən) *noun* A kind of asexual reproduction in which a whole animal develops from a part of the original animal.

a	cat	e	net	îr	gear	u	cup	u̇	look, pull	*th*	**this**	ə	alive,
ā	day, lake	ē	seed	o	hot	ū	fuse	oi	soil	hw	**wheel**		comet,
ä	father	i	fit	ō	cold	ûr	fur, bird	ou	**out**	zh	**measure**		acid, atom,
âr	dare	ī	pine	ô	paw	ü	tool, rule	th	**thin**	ŋ	**wing**		focus

rel·a·tive age (rel' ə tiv āj) *noun* The age of a rock as compared with another rock. See also *absolute age.*

relative dat·ing (rel' ə tiv dā' tiŋ) *noun* Putting artifacts or rock layers in chronological sequence as a way to compare their ages.

relative hu·mid·i·ty (rel' ə tiv hū mid' i tē) *noun* The amount of water vapor in the air compared with the maximum amount possible at that temperature and pressure, usually expressed as a percentage.

re·new·a·ble re·source (rē nü' ə bəl rē' sôrs) *noun* A resource, especially an energy source, that can be easily replaced or can never be used up, such as the wind, waves, or Sun.

re·pel (ri pel') *verb* To push away. *The north pole of one magnet repels the north pole of another magnet.*

rep·li·ca·tion (rep li kā' shən) *noun* The performance of an experiment more than once, following identical procedures each time. **replicate** *verb*

re·pro·duc·tion (rē prə duk' shən) *noun* The process by which a living thing creates more of its own kind. **reproduce** *verb*

rep·tile (rep' tīl) *noun* A cold-blooded animal that has a dry, protective covering of scales or plates and that lays eggs with leathery shells. *Alligators, snakes, and turtles are examples of reptiles.*

res·er·voir (rez' ər vwär) *noun* A natural or artificial storage area for supplies of fresh water.

re·sis·tance (ri zis' təns) *noun* 1. The force that a simple machine acts against. *The resistance in a simple machine is the load.* 2. The ability of certain materials to oppose the flow of electric current. *A substance with high resistance would not be a good conductor.*

resistance force (ri zis' təns fôrs) *noun* The force that a load exerts against an effort force.

re·sis·tor (ri zis' tər) *noun* A material that resists the flow of electric current. See also *insulator. Wire light bulb filaments that heat up and glow are examples of resistors.*

res·o·nance (rez' ə nəns) *noun* An increase in the size of vibrations caused by applying a force at an object's natural frequency. *Pushing a swing in the same rhythm it is already swinging is an example of resonance.* **resonate** *verb*

re·source (rē' sôrs) *noun* Something valuable or useful to a place or person. *Clean air and water are valuable resources for health.*

res·pi·ra·tion (res pə rā' shən) *noun* The process by which living things take in oxygen and give off carbon dioxide.

res·pi·ra·to·ry sys·tem (res' pər ə tôr ē sis' təm) *noun* The organ system that brings oxygen to body cells and removes waste gas. *The nose, mouth, windpipe, bronchial tubes, lungs, and diaphragm are all part of the respiratory system.*

re·sponse (ri spons') *noun* The way a living thing acts or behaves when it receives a stimulus.

ret·i·na (ret' ə nə) *noun* The layer of sensory cells that lines the back of the eyeball.

re·us·a·ble re·source (rē ūz′ ə bəl rē′ sôrs) *noun* Material or energy that can be used again.

re·use (rē ūz′) *verb* To use again. *One way to recycle objects is to reuse them.*

rev·o·lu·tion (rev ə lü′ shən) *noun* The movement of one object around another in an orbit.

re·volve (ri vôlv′) *verb* To move in a circle around a central point or object.

rhi·zoid (rī′ zoid) *adjective* Rootlike. *Some plants, such as mosses, produce rhizoid filaments.*

rhi·zome (ri′ zōm) *noun* A horizontal, underground plant stem that produces leaves and flowers above the soil surface and roots below. *Ferns and irises are examples of rhizomes.*

rib (rib) *noun* One of the curved bones in the chest that protect the heart and lungs.

Rich·ter scale (rik′ tər skāl) *noun* A graded measure from 1 to 10 of the energy released by an earthquake. *An earthquake with a 5.3 magnitude on the Richter scale is a moderate earthquake.*

ridge (rij) *noun* A long, narrow chain of mountains or hills, on land or in the ocean.

rift·ing (rif′ tiŋ) *noun* The pulling or pushing apart of two pieces of Earth's crust. *Rifting causes land between the sections of crust to drop and form deep valleys.*

rift vol·ca·no (rift vol kā′ nō) *noun* A volcano formed along a rift where two pieces of crust separate and magma from below wells up to fill the space.

Ring of Fire (riŋ əv fīr) *noun* A zone of frequent earthquakes and volcanic eruptions around the rim of the Pacific Ocean.

riv·er (riv′ ər) *noun* A large, natural stream of fresh water that flows into a lake or ocean.

rock (rok) *noun* The hard, mineral material that makes up Earth's crust. *The three main kinds of rocks are igneous, sedimentary, and metamorphic.*

rock cy·cle (rok sī′ kəl) *noun* An ongoing natural process by which rocks are changed from one type into another over long periods of time.

rock·et en·gine (rok′ it en′ jin) *noun* A powerful engine that thrusts a vehicle forward by burning fuel and oxygen and sending gases out the back.

root (rüt) *noun* The part of a plant that grows into the ground, holding the plant in place and absorbing water and minerals from the soil.

root cap (rüt kap) *noun* A mass of cells that covers and protects the root tip, where new root tissue is created.

ro·tate (rō′ tāt) *verb* To spin or turn around a central point.

a	cat	e	net	îr	gear	u	cup	u̇	look, pull	*th*	this	ə	alive,
ā	day, lake	ē	seed	o	hot	ū	fuse	oi	soil	hw	wheel		comet,
ä	father	i	fit	ō	cold	ûr	fur, bird	ou	out	zh	measure		acid, atom,
âr	dare	ī	pine	ô	paw	ü	tool, rule	th	thin	ŋ	wing		focus

ro·ta·tion (ro tā′ shən) *noun* The act of spinning or turning around a central point.

rud·der (rud′ ər) *noun* A hinged flap on the tail of an airplane or the back of a boat, used for steering.

run·off (run′ of) *noun* Water from rain or melted snow that flows along Earth's surface into bodies of water.

Ss

sa·lin·i·ty (sə lin′ i tē) *noun* The amount of salt in a solution. **saline** *adjective*

sa·li·va (sə lī′ və) *noun* The liquid in the mouth that keeps the mouth moist, softens food, and helps start digestion. **salivate** *verb*

sal·i·var·y gland (sal′ ə ver ē gland) *noun* One of the organs in the mouth that secretes saliva.

salt (sôlt) *noun* 1. A white mineral made of crystals used to season or preserve food. 2. A compound formed from the chemical reaction of an acid and a base.
Formula: NaCl

sap (sap) *noun* The liquid that circulates nutrients throughout a plant.

sap·ling (sap′ liŋ) *noun* A young tree.

sat·el·lite (sat′ ə līt) *noun* An object that revolves around a larger object, usually a heavenly body. *Our planet's moon is a natural satellite, but many artificial satellites also orbit Earth.*

Sat·urn (sat′ ərn) *noun* The planet that is the second largest in our solar system, sixth in distance from the Sun, and best known for its rings.

sa·van·na (sə van′ ə) *noun* A flat, grassy, tropical or subtropical plain that has few trees.

scale (skāl) *noun* 1. An instrument used to weigh objects. 2. One of the hard, thin plates of skin that covers some fish and reptiles. 3. A graduated measurement scheme, such as a ruler or the Beaufort scale for describing wind. 4. The relationship between two sets of measurements, such as a map scale of 1 inch equaling 1 mile.

scap·u·la (skap′ yə lə) *noun, singular* One of the two shoulder blades.
scapulae, scapulas *plural*

scav·en·ger (skav′ ən jər) *noun* An animal that eats dead animals or garbage.

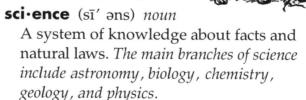

sci·ence (sī′ əns) *noun* A system of knowledge about facts and natural laws. *The main branches of science include astronomy, biology, chemistry, geology, and physics.*

sci·en·tif·ic meth·od (sī ən tif′ ik meth′ əd) *noun* A step-by-step procedure often used to solve a problem or discover why things happen. *The scientific method involves making observations, forming hypotheses, doing experiments, and drawing conclusions.*

scientific name (sī ən tif' ik nām) *noun*
The names of an organism's genus and
species. Homo sapiens *is the scientific name
for humans.*

sci·en·tist (sī' ən tist) *noun* A person who
is educated and trained in one of the
fields of scientific study.

scler·a (skler' ə) *noun* The white outer
coating that covers all of the eyeball
except the cornea. **scleral** *adjective*

screw (skrü) *noun* A simple machine made
up of an inclined plane wrapped around a
center post, creating threads.

scu·ba (scü' bə) *noun* A portable supply
of compressed air that allows a swimmer
or diver to breathe underwater. Stands for
Self-Contained Underwater Breathing
Apparatus.

sea breeze (sē brēz) *noun* A gentle,
cooling wind blowing off the ocean onto
land.

sea-floor spread·ing (sē' flôr spred' iŋ)
noun The process by which magma
creates new ocean crust at the midocean
ridges, expanding the sea bottom.

sea·mount (sē' mount) *noun* A usually
submerged mountain rising from the
ocean floor. *Some seamounts rise above the
water, and we see them as islands.*

sec·ond·ar·y con·sum·er (sek' ən der ē
kən sü' mər) *noun* An organism that eats a
primary consumer, such as a hawk that
eats a rabbit. See also *primary consumer.*

secondary wave (sek' ən der ē wāv)
noun An earthquake wave that travels
through rock and vibrates at right angles
to the direction of travel.
Abbreviation: S wave.

Sec·ond Law of Mo·tion (sek' ənd lô
əv mō' shən) *noun* Isaac Newton's law
stating that acceleration is produced when
a force acts on a mass, and the greater the
mass of an object, the greater the force
needed to accelerate it.

sed·i·ment (sed' ə mənt) *noun* Matter that
settles to the bottom of a liquid, especially
materials such as sand and gravel
deposited by wind, water, and ice in
bodies of water.

sed·i·men·ta·ry rock (sed ə mən' tə rē
rok) *noun* A type of rock formed when
layers of sediment are pressed and
cemented together over time.

seed (sēd) *noun*
The part of a flowering
plant that can develop
into a new plant. *A
seed includes an embryo,
food tissue, and a
protective coat.*

seed coat (sēd kōt) *noun* The outer
protective covering of a seed.

seed dis·per·sal (sēd dis pûr' səl) *noun*
The movement of seeds away from the
parent plant.

seed leaf (sēd lēf) *noun* Part of the
embryo in a seed plant; also called a
cotyledon. See also *dicot, monocot.*

a	cat	e	net	îr	gear	u	cup	ù	look, pull	*th*	this	ə	alive,
ā	day, lake	ē	seed	o	hot	ū	fuse	oi	soil	hw	wheel		comet,
ä	father	i	fit	ō	cold	ûr	fur, bird	ou	out	zh	measure		acid, atom,
âr	dare	ī	pine	ô	paw	ü	tool, rule	th	thin	ŋ	wing		focus

seed·ling (sēd′ liŋ) *noun* A young plant grown from a seed, not a cutting.

seed plant (sēd plant) *noun* A plant that produces and grows from seeds.

seis·mic wave (sīz′ mik wāv) *noun* A vibration that radiates out from the focus of an earthquake.

seis·mo·graph (sīz′ mə graf) *noun* An instrument that measures and records the time, strength, and duration of earthquake vibrations.

seis·mom·e·ter (sīz mom′ ə tər) *noun* A seismograph that measures the actual movements of the ground during an earthquake.

se·lec·tive breed·ing (sə lek′ tiv brēd′ iŋ) *noun* Crossing carefully chosen plants or animals to produce offspring with certain desirable traits.

self-pol·li·na·tion (self pol ə nā′ shən) *noun* The transfer of pollen from the male part (anther) to the female part (stigma) of the same flower. See also *cross-pollination*.

sem·i·cir·cu·lar ca·nal (sem ē sûr′ kyə lər kə nal′) *noun* One of three loop-shaped parts of the inner ear that helps maintain body balance.

sense or·gan (sens ôr′ gən) *noun* A body structure that takes in information from the environment—in humans, the eyes, ears, nose, taste buds, and skin.

sen·so·ry neu·ron (sen′ sə rē nûr′ on) *noun* A nerve cell that receives input from sensory cells.

se·pal (sē′ pəl) *noun* One of the leaves that surround and protect a flower bud, open to allow it to bloom, then remain at the base of the flower.

se·quenc·ing (sē′ kwən siŋ) *verb*
1. Arranging in a regular order.
2. Deciding on the order of chemical components.

se·ries cir·cuit (sîr′ ēz sûr′ kit) *noun* An electric circuit in which all the devices are connected one after the other so that the current flows in a single path.

set·tling (set′ liŋ) *verb* Sinking gradually to the bottom of a liquid.

sex cell (seks sel) *noun* A male or female reproductive cell.

sex-linked gene (seks′ liŋkt jēn) *noun* A gene located on the chromosome that determines the individual's sex or gender.

sex·u·al re·pro·duc·tion (sek′ shü əl rē prə duk′ shən) *noun* The process by which cells from two parents unite to produce an offspring. See also *asexual reproduction*.

shear (shîr) *noun* A movement of plates that pushes one part of Earth's crust past another.

shield vol·ca·no (shēld vol kā′ nō) *noun* A wide, rounded volcano formed by very fluid lava over time.

shore (shôr) *noun* Land bordering a large body of water, such as the ocean.

shore·line (shôr līn) *noun* The boundary where a body of water meets the land.

short cir·cuit (shôrt sûr′ kit) *noun* A surge of electric current through a circuit because of a sudden drop in resistance. *In a short circuit, the current does not complete the full circuit but jumps to the ground or the end point, and any connected appliance will stop working.*

short-day plant (shôrt′ dā plant) *noun* A plant that blooms when there is more darkness than daylight.

sick·le-cell a·ne·mia (sik′ əl sel ə nē′ mē ə) *noun* An inherited disease marked by sickle-shaped red blood cells.

sill (sil) *noun* A horizontal layer of rock formed when magma hardens between other layers.

sil·ver (sil′ vər) *noun* An element that is a soft, white metal with high luster.
Symbol: Ag

sim·ple ma·chine (sim′ pəl mə shēn′) *noun* One of six basic devices that make it easier to do work—lever, pulley, wedge, screw, inclined plane, and wheel and axle.

simple mi·cro·scope (sim′ pəl mī′ krə skōp) *noun* A microscope with one lens between the object and the eye, usually with lower magnification than a compound microscope.

skel·e·tal mus·cle (skel′ i təl mus′ əl) *noun* A striated, voluntary muscle attached to a bone in order to cause movement.

skeletal sys·tem (skel′ i təl sis′ təm) *noun* In vertebrates, the body system that gives shape and structure to the body and is made up of bones, cartilage, ligaments, and joints.

skull (skul) *noun* In vertebrates, the bony framework of the head that encloses and protects the brain.

slope (slōp) 1. *noun* Land on an incline. 2. *verb* To slant up or down.

small in·tes·tine (smol in tes′ tin) *noun* The long, tubelike part of the digestive system between the stomach and large intestine where most nutrients from food are absorbed into the bloodstream.

smelt (smelt) *verb* To melt or fuse ore, usually to separate pure metals from impurities.

smog (smog) *noun* A kind of air pollution caused by a combination of smoke, chemicals, and fog.

smooth mus·cle (smü*th* mus′ əl) *noun* A type of muscle found in the lining of such organs as the stomach and bladder.

soil (soil) *noun* The top layer of Earth's surface in which plants grow, made up of sand, silt, clay, and humus.

so·lar cell (sō′ lər sel) *noun* A device that converts light energy from the Sun to electrical energy and is used as a power source.

solar e·clipse (sō′ lər i klips′) *noun* The temporary blocking of the Sun when the Moon is between Earth and the Sun, so that Earth passes through the Moon's shadow.

solar en·er·gy (sō′ lər en′ ər jē) *noun* Energy from the Sun.

solar flare (sō′ lər flâr) *noun* A sudden energy outburst on a small area of the Sun's surface.

a	cat	e	net	îr	gear	u	cup	u̇	look, pull	*th*	this	ə	alive,
ā	day, lake	ē	seed	o	hot	ū	fuse	oi	soil	hw	wheel		comet,
ä	father	i	fit	ō	cold	ûr	fur, bird	ou	out	zh	measure		acid, atom,
âr	dare	ī	pine	ô	paw	ü	tool, rule	th	thin	ŋ	wing		focus

so·lar sys·tem (sō′ lər sis′ təm) *noun*
A star and the celestial objects that orbit it.

so·lar wind (sō′ lər wind) *noun* A
constant stream of particles given off by
the Sun, moving at about 400 kilometers
(250 miles) per second.

sol·id (sol′ id) *noun* Matter that is not
liquid or gas and usually keeps a definite
size and shape. *Ice is the solid form of water.*

sol·stice (sôl′ stəs) *noun* Either of two
times of the year when the Sun is farthest
north or south of the equator, creating
days with the longest and shortest
amounts of daylight.

sol·u·bil·i·ty (sol yə bil′ i tē) *noun*
The ability of a substance to be dissolved
in another substance.

sol·u·ble (sol′ yə bəl) *adjective* Capable of
being dissolved in or by a liquid. See also
insoluble.

sol·ute (sol′ ūt) *noun* A substance that
dissolves in another substance to make a
solution. *Salt can be one of many solutes in
ocean water.*

so·lu·tion (sə lü′ shən) *noun* A mixture of a
dissolved substance and the liquid in
which it is dissolved, thoroughly blended
to be homogeneous. *Mixing chocolate drink
powder and milk makes a tasty solution.*

sol·vent (sol′ vənt) *noun* A substance,
usually a liquid, that can dissolve another
substance.

so·nar (sō′ när) *noun* A method or device
that detects and locates underwater
objects or water depth by bouncing sound
waves off the objects or bottom. Stands for
SOund Navigation And Ranging.

sound (sound) *noun* 1. Something that can
be heard. 2. A form of energy that is
transmitted by an object vibrating.

sound en·er·gy (sound en′ ər jē) *noun*
The difference between the total energy
and the energy that would exist if no
sound waves were present; also known as
acoustic energy.

sound syn·the·siz·er (sound
sin′ thə sī zər) *noun* A computerized
electronic device for producing and
controlling sound.

sound wave (sound wāv) *noun*
A vibration caused by a moving object
that passes through air, liquid, or solid
and can be heard.

South·ern Hem·i·sphere (su*th*′ ərn
hem′ i sfîr) *noun* The half of Earth that is
south of the equator.

space probe (spās prōb) *noun*
An uncrewed spacecraft sent to study
planets and other distant objects in our
solar system.

space shut·tle (spās shut′ əl) *noun*
A reusable spacecraft to transport people
and supplies between
Earth and space. *Space
shuttles must be powered
by rockets when launched,
but they can be flown and
landed like airplanes.*

spawn (spôn) *verb* To produce or deposit
eggs, especially in large numbers.

spe·cies (spē′ shēz) *noun* In classification,
the smallest grouping of organisms, made
up of one kind of plant or animal.

spec·i·men (spes′ ə mən) *noun* A sample
that represents its entire group, usually
for scientific study.

spec·tro·scope (spek′ trə skōp) *noun*
A device for forming and examining
spectra, especially those in the visible
range.

spec·trum (spek′ trəm) *noun, singular*
The band of colors produced when light
passes through a prism.
spectra, spectrums *plural*

speed (spēd) *noun* The rate of motion. *The
speed of motor vehicles is usually measured in
kilometers or miles per hour.*

spi·nal cord (spī′ nəl kôrd) *noun*
The bundle of nerve tissue that runs from
the brain through the spinal column,
carries signals between the brain and the
nerves, and controls many reflex actions.

spi·ral gal·ax·y (spī′ rəl gal′ ək sē) *noun*
A galaxy with curved arms of matter that
give it a spiral or whirlpool shape.

spore (spôr) *noun*
A plant cell that grows
into a new plant and is
produced by plants
that do not flower, such
as fungi, mosses, and
ferns.

sprain (sprān) *noun* A joint injury that
stretches or tears ligaments, usually
caused by twisting or wrenching.

spring (spriŋ) *noun* 1. The season between
winter and summer, officially from the
March equinox to the June solstice in the
Northern Hemisphere. 2. A place where
groundwater comes out of the ground.

spring tide (spriŋ tīd) *noun* A very high
tide that comes around the times of the
new Moon and full Moon phases. See also
neap tide.

sprout (sprout) *1. noun* A young plant
beginning to grow from a seed. *2. verb* To
begin to grow, as when a seed germinates
and produces a root and a shoot.

sta·bil·i·ty (stə bil′ i tē) *noun* Steadiness, or
resistance to change.

sta·men (stā′ mən) *noun* The male part of a
flower, which produces pollen.

stan·dard time zone (stan′ dərd tīm
zōn) *noun* One of the 24 zones into which
Earth is divided corresponding to the 24
hours in a day.

standard u·nit (stan′ dərd ū′ nit) *noun*
An amount or measurement agreed to
and used by many people.

star (stär) *noun* A large celestial ball of gas
that shines by its own light. *The nearest
star to Earth outside our solar system is
Alpha Centauri, 4 light-years away.*

starch (stärch) *noun* A food substance
without taste or smell that is found in
potatoes, grains, beans, and other
vegetables.

state of mat·ter (stāt əv mat′ ər) *noun*
One of the three forms in which matter
can exist—solid, liquid, or gas.

stat·ic e·lec·tric·i·ty
(stat′ ik i lek tris′ i tē) *noun* An electric
charge that builds up in an object, such as
from friction, and stays there.

sta·tion·ary front (stā′ shən âr ē frunt)
noun The boundary between a cool and
a warm air mass, neither of which is
moving.

a	cat	e	net	îr	gear	u	cup	u̇	look, pull	*th*	this	ə	alive,
ā	day, lake	ē	seed	o	hot	ū	fuse	oi	soil		hw	wheel		comet,
ä	father	i	fit	ō	cold	ûr	fur, bird	ou	out		zh	measure		acid, atom,
âr	dare	ī	pine	ô	paw	ü	tool, rule	th	thin		ŋ	wing		focus

sta·tis·ti·cal fore·cast·ing (stə tis′ tə kəl fôr′ kas tiŋ) *noun* The prediction of weather based on the probability that past weather patterns will repeat.

steam en·gine (stēm en′ jin) *noun* An engine powered by steam, which is produced when water is boiled by the burning of fuel.

stem (stem) *noun* The part of a plant that grows up out of the ground and supports leaves, flowers, and fruit.

ster·num (stûr′ nəm) *noun, singular* The breastbone. **sternums, sterna** *plural*

ste·roid (ster′ oid) *noun* A chemical substance found naturally in plants and animals that is important for good health but can be abused if taken improperly.

stim·u·lant (stim′ yə lənt) *noun* A substance that speeds up the activity of some body part. *Caffeine is a stimulant found in coffee, tea, and the kola nut.*

stim·u·lus (stim′ yə ləs) *noun, singular* Something that causes a reaction in an organism. **stimuli** *plural*

sto·ma (stō′ mə) *noun, singular* A tiny opening on a leaf's surface through which gases pass. **stomata** *plural*

stom·ach (stum′ ək) *noun* The pouchlike organ in the body where food is turned into mush by acids, digestive enzymes, and muscle activity.

storm surge (stôrm sûrj) *noun* A great rise of water pushed to a coastline by a powerful storm.

strain (strān) *1. noun* A muscle injury caused by excessive tension, effort, or use. *2. verb* To pass through a sieve to separate solids from liquids. 3. To draw or pull tight. 4. To exert maximum effort.

strat·o·sphere (strat′ ə sfîr) *noun* The layer of Earth's atmosphere above the troposphere, from about 11 to 50 kilometers (7 to 31 miles) above Earth's surface.

stra·tus cloud (strat′ əs kloud) *noun* A gray, layered cloud that forms at low altitudes of about 600 to 2,100 meters (2,000 to 7,000 feet) and usually brings precipitation.

streak (strēk) *noun* The fine deposit of dust, especially its color, left on an abrasive surface when a mineral is scraped across it. *Streak is more reliable than the visible color of the specimen in identifying a mineral.*

streak plate (strēk plāt) *noun* A ceramic abrasive surface for streak tests.

stream·line (strēm′ līn) *verb* To design or make in a way that reduces drag.

stress (stres) *noun* 1. A response to pressure. 2. The effect of high heat or pressure on rocks.

stri·at·ed mus·cle (strī′ ā təd mus′ əl) *noun* Skeletal muscle tissue marked by dark and light bands that look like stripes.

strip farm·ing (strip färm′ iŋ) *noun* A farming method in which widely spaced plants are alternated with tightly growing grasses, used to save water runoff.

struc·tur·al ad·ap·ta·tion (struk′ chər əl ad ap tā′ shən) *noun* A change that occurs over time in an animal's structure to help it survive and reproduce in an environment.

struc·ture (struk′ chər) *noun* The way in which something is organized or put together. *The structure of a plant cell is quite different from the structure of an animal cell.*

sub·a·tom·ic (sub ə tom′ ik) *adjective*
Describes a particle smaller than an atom or inside an atom.

sub·duc·tion (sub duk′ shən) *noun*
The action of the edge of one section of Earth's crust sliding below the edge of another section.

sub·mers·i·ble (sub mûr′ sə bəl) *adjective*
Capable of being underwater or of working underwater.

sub·soil (sub′ soil) *noun* The layer of soil underneath the surface layer.

sub·stance (sub′ stəns) *noun* Physical material that has weight and takes up space. *The page you are now reading is a substance.*

suc·ces·sion (sək sesh′ ən) *noun*
The natural series of changes in the makeup of an ecosystem or a species over time.

su·crose (sü′ krōs) *noun* A sugar found in most plants, taken from sugarcane and sugar beets for human use.

sug·ar (shug′ ər) *noun* A sweet, crystalline carbohydrate used in foods and drinks.

sun·spot (sun′ spot) *noun, plural* A dark spot on the Sun's surface caused by a magnetic storm that cools a large area.

su·per·no·va (sü pər nō′ və) *noun*
A massive exploding star that can give off a billion times more light than the Sun.

su·per·po·si·tion (sü pər pə zish′ ən) *noun*
The idea that in layers of rock, the top layers are the youngest and the bottom layers are the oldest.

su·per·son·ic (sü pər son′ ik) *adjective*
Moving faster than the speed of sound. See also *hypersonic*.

sur·face cur·rent (sûr′ fəs kûr′ ənt) *noun*
An ocean current caused by the friction of prevailing winds over the water's surface.

surface ten·sion (sûr′ fəs ten′ shən) *noun*
The attraction of molecules in a liquid to one another that produces a skinlike covering on the surface of the liquid. *Surface tension can enable you to float a pin on water in a glass.*

surface wave (sûr′ fəs wāv) *noun* One of the wavelike vibrations during an earthquake that cause much of the structural damage. **Abbreviation: S wave**

sur·vey·or (sûr vā′ ər) *noun* A person who measures and records the size and shape of plots of land. **survey** *verb*

sus·pen·sion (sə spen′ shən) *noun*
A mixture in which solid particles hang within the liquid. See also *colloid, emulsion*.

switch (swich) *noun* A device for opening or closing an electrical circuit. *When you flip a switch to the "on" position, the circuit is completed and an attached bulb will light.*

sym·bi·o·sis (sim bē ō′ sis) *noun* A close relationship of two unlike organisms.

sym·bol (sim′ bəl) *noun* An object or a picture that stands for something else.

a	cat	e	net	îr	gear	u	cup	ů	look, pull	*th*	this	ə	alive,
ā	day, lake	ē	seed	o	hot	ū	fuse	oi	soil	hw	wheel		comet,
ä	father	i	fit	ō	cold	ûr	fur, bird	ou	out	zh	measure		acid, atom,
âr	dare	ī	pine	ô	paw	ü	tool, rule	th	thin	ŋ	wing		focus

sym·met·ri·cal (si mət′ rik əl) *noun*
Balanced or having an equal arrangement of parts on the two sides of a center line.

syn·apse (sin′ aps) *noun* The point where impulses pass from one nerve cell to another. *The synapse is the place where the dendrite of one nerve cell meets the axon of another.* **synaptic** *adjective*

syn·op·tic weath·er map (sin op′ tik weth′ ər map) *noun* A map that shows weather conditions over a large area.

sys·tem (sis′ təm) *noun* Two or more objects that work together in a meaningful way. *The body is made up of several organ systems.*

tap·root (tap′ rüt) *noun* A main root that grows straight downward and gives off smaller roots.

taste bud (tāst bud) *noun* A cluster of cells, mainly in the tongue, that senses whether something is sweet, sour, salty, or bitter.

tec·ton·ic plate (tek ton′ ik plāt) *noun* One of the large, rigid pieces of rock that make up Earth's crust.

tel·e·graph (tel′ ə graf) *noun* A device that sends coded signals over distance, especially by electrical signals over wires.

tel·e·phone (tel′ ə fōn) *noun* An instrument for reproducing sounds at a distance, specifically one that sends electrical impulses over wires or by radio waves.

tel·e·scope (tel′ ə skōp) *noun* An instrument for viewing distant objects in space that uses lenses or mirrors to make the objects appear closer and larger.

tem·per·ate (tem′ pər ət) *adjective* Having neither a very cold nor a very hot climate.

tem·per·a·ture (tem′ pər ə chər) *noun* A measure of how hot or cold something is.

ten·don (ten′ dən) *noun* A tough, strong, thick cord or band of tissue that joins a muscle to a bone or other body part.

ten·sion (ten′ shən) *noun* The tightness or tautness of a material produced by the pull of opposing forces. *The tension on the tug-of-war rope suddenly increased when the referee shouted, "Go!"*

ter·mi·nal (tûr′ mə nəl) 1. *noun* One of two places on an electric cell where current leaves or enters the power source. *A battery has a positive and a negative terminal.* 2. *adjective* Describes an end point or boundary. *The terminal bud on a branch is often the latest to have developed.*

ter·rac·ing (ter′ is iŋ) *verb* Making stepped ridges on a hillside to increase the area that can be farmed or to reduce runoff or erosion. **terrace** *noun*

ter·rar·i·um (tə râr′ ē əm) *noun* A closed, clear container for raising and observing plants and small animals indoors.

ter·res·tri·al plan·et (tə res′ trē əl plan′ it) *noun* A planet that is Earthlike and mostly rock, unlike those composed of gas.

tex·ture (teks' chər) *noun* How a material feels to the touch. *The texture of a rock can be described using words such as coarse, fine, glassy, angular, or rounded.*

the·o·ry (thē' ə rē) *noun* An idea or statement of how or why something happens.

theory of con·ti·nen·tal drift (thē' ə rē əv kon ti nen' təl drift) *noun* An early concept that the continents are not stationary but moved to their present locations.

theory of ev·o·lu·tion (thē' ə rē əv ev ə lü' shən) *noun* The theory that animals and plants now on Earth developed by a gradual and continuous change from previously existing life forms.

theory of plate tec·ton·ics (thē' ə rē əv plāt tek ton' iks) *noun* A theory that Earth is divided into a small number of massive plates that float independently on the crust and are constantly in very slow motion over Earth's surface.

ther·mal en·er·gy (thûr' məl en' ər jē) *noun* The kinetic energy of the moving particles of an object.

thermal ex·pan·sion (thûr' məl ek span' shən) *noun* The spreading of matter as its temperature rises.

thermal pol·lu·tion (thûr' məl pə lü' shən) *noun* The release of excess heat into the air or water at a temperature harmful to the environment.

ther·mom·e·ter (thûr mom' ə tər) *noun* An instrument used to measure temperature.

Third Law of Mo·tion (thûrd lô əv mō' shən) *noun* Isaac Newton's law stating that for every action there is a reaction equal in strength and opposite in direction.

third quar·ter Moon (thûrd kwôr' tər mün) *noun* The Moon phase when the left half of the Moon is visible and growing smaller; also called last quarter.

thread (thred) *noun* The ridge or groove that spirals around a screw. *The thread of a wood screw is designed to dig into the wood.*

three-di·men·sion·al (thrē di men' shən əl) *adjective* Describes an object that has three dimensions: length, width, and depth.

thrive (thrīv) *verb* To live and grow in a strong, healthy way.

thrust (thrust) *noun* A force that moves an object forward, especially the force produced by a propeller or an engine that drives a vehicle or an aircraft.

thun·der (thun' dər) *noun* The loud rumbling or cracking sound that follows a flash of lightning, caused by the sudden expansion of air in the path of the electrical discharge.

thun·der·head (thun' dər hed) *noun* The cumulonimbus cloud in which a thunderstorm forms.

a	cat	e	net	îr	gear	u	cup	u̇	look, pull	*th*	**this**	ə	alive,
ā	day, lake	ē	seed	o	hot	ū	fuse	oi	soil	hw	**wheel**		comet,
ä	father	i	fit	ō	cold	ûr	fur, bird	ou	out	zh	measure		acid, atom,
âr	dare	ī	pine	ô	paw	ü	tool, rule	th	thin	ŋ	wing		focus

thun·der·storm (thun′ dər stôrm)
noun A storm with thunder, lightning, and usually heavy rain.

tib·i·a (tib′ ē ə) *noun* The inner and larger of the two bones in the lower leg, between the knee and ankle. See also *fibula.*

tid·al en·er·gy (tī′ dəl en′ ər jē) *noun* The energy produced by tide changes and wave action, particularly as it may be harnessed for human use.

tidal pow·er plant (tī′ dəl pou′ ər plant) *noun* A building with turbines for changing the energy of ocean tides into electric power, often a dam built across an estuary.

tide (tīd) *noun* The regular, alternating rise and fall of the ocean's surface level, caused by the gravitational pull of the Moon and Sun on Earth.

tide pool (tīd pül) *noun* A small body of water remaining on the land after the tide has gone out, or ebbed; also called tidal pool.

till (til) *noun* A jumble of different sizes of sediment deposited by a glacier, including clay, sand, gravel, and boulders.

tim·bre (tim′ bər) *noun* The quality given to a sound by its overtones. *I recognized his voice by its timbre.*

time (tīm) *noun* 1. A measure of the past, present, and future based on natural events such as seasons. 2. A system of measuring the passing of seconds, minutes, hours, and so on. *Clocks and watches are the major instruments used to measure and display time.*

tis·sue (tish′ ü) *noun* A group of similar cells that form one of an organism's structures.

tissue cul·ture (tish′ ü kul′ chər) *noun* The process of making body tissue grow in a medium outside the organism.

top·soil (top′ soil) *noun* The surface layer of soil, where most plants grow.

tor·na·do (tôr nā′ dō) *noun* A violent whirling wind with a funnel-shaped cloud that moves in a narrow destructive path over the ground.

tox·in (tok′ sin) *noun* A poisonous substance that is a product of a living organism. **toxic** *adjective*

tra·chea (trā′ kē ə) *noun* The tube through which air passes between the nose and mouth and the lungs. Also called the *windpipe.*

trac·ing (trā′ siŋ) *noun* The line drawn by an instrument such as a seismograph that records movement.

trade wind (trād wind) *noun* One of the winds in a belt around Earth that move from high-pressure zones toward the low pressure at the equator.

trait (trāt) *noun* A quality or characteristic that makes one organism different from another. *You can see many different traits when you compare a fish and an elephant.*

trans·form·er (trans fôr′ mər) *noun* A device that changes the voltage of an electric current. *A step-down transformer reduces voltage to safe levels before the current enters homes.*

trans·form fault (trans′ fôrm fôlt) *noun* A boundary where tectonic plates slide past each other.

trans·fu·sion (trans fū′ zhən) *noun*
The transfer of blood or plasma from one person's body to another's for medical purposes.

trans·lu·cent (trans lü′ sənt) *adjective*
Letting some light through but not clear or transparent.

trans·mit (trans mit′) *verb* To send or carry from one person or place to another.
transmission *noun*

trans·par·ent (trans pâr′ ənt) *adjective*
Letting light through so that objects on the other side can be seen clearly.

tran·spi·ra·tion (trans pə rā′ shən) *noun*
The release of water vapor into the air by plants through their stomata.

trench (trench) *noun* A deep canyon on the ocean floor where subduction occurs.

trib·u·tary (trib′ yə ter ē) *noun* A stream or river that feeds a larger one.

tri·ceps (trī′ seps) *noun*
The muscle in the back of the upper arm that extends or straightens the lower arm when it contracts.

trop·i·cal (trop′ i kəl) *adjective* Describes a region or climate that is frost free and supports year-round plant growth.

tro·pism (trō′ piz əm) *noun* Growth of a plant toward or away from a stimulant.

tro·po·sphere (trop′ ə sfîr) *noun*
The lowest layer of Earth's atmosphere, where weather occurs.

trough (trof) *noun* 1. A long, narrow, shallow depression between hills on land or on the sea floor. 2. The lowest point in a wave. *Wave height in the ocean is measured from the trough to the crest.*

tsu·na·mi (tsü nä′ mē) *noun* A huge ocean wave caused by an underwater earthquake or a volcanic eruption. Also called *tidal wave.*

tu·ber (tü′ bər) *noun*
The thick or swollen underground stem of a plant, such as the potato.

tun·dra (tun′ drə) *noun*
A flat or slightly rolling plain with few trees and permanently frozen ground.

tune (tün) *verb* To adjust precisely, as in musical pitch. *She tuned the guitar before she began to play the song.*

tur·bid·i·ty cur·rent (tûr bid′ i tē kur′ ənt) *noun* Rapidly moving, sediment-filled water that flows down a slope into a larger body of water.

tur·bine (tûr′ bən, tûr′ bīn) *noun*
An engine in which water, steam, or gas (including wind) passes through the blades of a wheel, making it revolve.

ty·phoon (tī fün′) *noun* A tropical cyclone occurring in the western Pacific Ocean.

a	cat	e	net	îr	gear	u	cup	u̇	look, pull	*th*	this	ə	alive,
ā	day, lake	ē	seed	o	hot	ū	fuse	oi	soil		hw	wheel		comet,
ä	father	i	fit	ō	cold	ûr	fur, bird	ou	out	zh	measure		acid, atom,
âr	dare	ī	pine	ô	paw	ü	tool, rule	th	thin	ŋ	wing		focus

ul·na (ulʹ nə) *noun* The bone on the little-finger side of the forearm.

ul·tra·son·ic (ul trə sonʹ ik) *adjective* Having a sound too high in frequency to be heard by the human ear.

ul·tra·vi·o·let (ul trə vīʹ ə lit) *adjective* Just beyond the violet end of the visible spectrum and thus invisible to the human eye.

um·bil·i·cal cord (um bilʹ i kəl kôrd) *noun* The tube that connects a fetus to its mother's body, through which it gets its oxygen and nourishment.

un·bal·anced for·ces (un balʹ ənst fôrsʹ əz) *noun, plural* Forces that do not cancel each other out when acting on a single object.

u·ni·verse (ūʹ ni vûrs) *noun* The billions of galaxies and the space in which they exist.

up·draft (upʹ draft) *noun* An upward movement of any gas, especially a sudden rush of heated air during a thunderstorm.

up·per man·tle (upʹ ər manʹ təl) *noun* The upper part of the layer of Earth between the crust and the core.

up·well·ing (upʹ wel iŋ) *noun* The upward movement of deeper water to the surface, especially along some shores.

Ur·a·nus (yûrʹ ə nəs) *noun* The seventh planet from the Sun in our solar system.

u·re·a (yù rēʹ ə) *noun* The organic compound that is the solid part of urine.

u·re·ter (yùʹ rə tər, yù rēʹ tər) *noun* One of two tubes that carries urine from the kidneys to the bladder.

u·re·thra (yù rēthʹ rə) *noun* The tube that carries urine from the bladder out of the body.

u·ri·na·ry blad·der (yûrʹ i ner ē bladʹ er) *noun* The sac that holds urine until it is removed from the body.

u·rine (yûrʹ in) *noun* Liquid waste produced by humans and other vertebrates.

u·ter·us (ūʹ tər əs) *noun* The hollow organ in female mammals in which unborn offspring grow; also called *womb*.

vac·ci·na·tion (vak si nāʹ shən) *noun* Protection against a disease provided by dead or weakened germs that are injected or swallowed; also called immunization.

vac·cine (vak sēnʹ) *noun* A substance made of dead or weakened germs introduced into the body to protect against a particular disease. *Vaccines cause the body to produce antibodies that fight disease.* **vaccinate** *verb*

vac·u·ole (vakʹ ū ōl) *noun* A small sac or space in the tissues of an organism that contains air or fluid.

vac·u·um (vakʹ ū əm) *noun* A space empty of air or matter.

val·ley (val' ē) *noun*
A long, low area of land between hills or ranges of hills or mountains.

valley breeze (val' ē brēz) *noun* A cool wind that blows up a mountain slope and replaces the slope's rising, sun-warmed air.

valve (valv) *noun* A device that opens and closes like a door to control a flow, such as a flow of blood. *Valves in the heart separate the chambers that receive blood from the chambers that pump blood.*

va·por·i·za·tion (vā pər i zā' shən) *noun* The process of changing a liquid or a solid to a gas by heating.

var·i·a·ble (vâr' ē ə bəl) *noun* The condition or factor that changes in an experiment so the effects can be observed. See also *control.*

var·i·a·tion (vâr ē ā' shən) *noun* 1. A difference from the usual or average. 2. A small difference between similar objects.

vas·cu·lar (vas' kyə lər) *adjective* Having tubes that carry bodily fluids, such as the blood of an animal or the sap of a plant.

vascular plant (vas' kyə lər plant) *noun* A plant with an internal system of tubes for transporting water, minerals, and food to its roots, stems, and leaves.

vascular sys·tem (vas' kyə lər sis' təm) *noun* The system of vessels that carry fluids throughout an organism.

veg·e·ta·bles (vej' tə bəlz) *noun, plural* Foods that come from plant parts such as seeds (corn), leaves (lettuce), roots (carrots), fruits (tomatoes), bulbs (onions), stems (broccoli), or tubers (potatoes).

veg·e·ta·tive prop·a·ga·tion (vej i tā' tiv prop ə gā' shən) *noun* The ability of plants to reproduce from their own structures, seeds, or spores, such as by sending out runners.

vein (vān) *noun* 1. A blood vessel that carries blood toward the heart. 2. The tubelike structure of a leaf. 3. A narrow channel in rock filled with minerals, especially metallic ores. 4. One of the fine ribs that form the framework of an insect's wings.

ve·loc·i·ty (və los' i tē) *noun* The rate at which an object moves in a certain direction. *The rocket's velocity increases as it travels toward space.*

ven·om (ven' əm) *noun* A poison produced by certain animals.

vent (vent) 1. *noun* The weak area in rock through which magma can escape. *The main opening in a volcano is a vent.* 2. *verb* To provide with a vent or to expel through a vent. *When working with certain chemicals, you should vent the room by opening a window.*

a	cat	e	net	îr	gear	u	cup	ú	look, pull	*th*	**this**	ə	alive,
ā	day, lake	ē	seed	o	hot	ū	fuse	oi	soil	hw	**wheel**		comet,
ä	father	i	fit	ō	cold	ûr	**fur**, bird	ou	**out**	zh	measure		acid, atom,
âr	dare	ī	pine	ô	paw	ü	**tool**, rule	th	**thin**	ŋ	wi**ng**		focus

ven·tri·cle (ven′ tri kəl) *noun* One of the two lower chambers of the heart, which pump blood out to the lungs and the rest of the body.

Ve·nus (vē′ nəs) *noun* The second planet from the Sun in our solar system.

ver·te·bra (vûr′ tə brə) *noun, singular* One of the bones that make up the spinal column, or backbone.
vertebrae, vertebras *plural*

ver·te·brate (vûr′ tə brāt) *noun* An animal with a spinal column, or backbone. See also *invertebrate*.

vi·bra·tion (vī brā′ shən) *noun* A rapid back-and-forth motion.

vil·lus (vil′ əs) *noun, singular* One of the tiny fingerlike structures that line the small intestine and allow nutrients to pass from the digestive system into the blood.
villi *plural*

vi·ral (vī′ rəl) *adjective* Caused by a virus.

vi·rus (vī′ rəs) *noun* A disease-producing agent smaller than a bacterium.

vis·i·ble light (viz′ ə bəl līt) *noun* The spectrum of colors that can be seen by the human eye.

visible spec·trum (viz′ ə bəl spek′ trəm) *noun* The bands of colors that become visible when light is refracted.

vi·ta·min (vī′ tə min) *noun* A nutrient found in food or produced by the body that is needed for good health.

vol·ca·no (vol kā′ nō) *noun* An opening in Earth's crust through which lava, ash, and cinders erupt, or the mountain formed from past eruptions.

volt (vōlt) *noun* A unit for measuring the force of an electric current.

volt·age (vōl′ tij) *noun* The force of an electric current, expressed in volts.

volt·me·ter (vōlt′ mē tər) *noun* A tool that measures voltage between two points in an electric circuit.

vol·ume (vol′ ūm) *noun* 1. The measure of the amount of space a three-dimensional object occupies. *A 100 mL glass can hold a greater volume than a 40 mL glass.* 2. The loudness of a sound, determined by the height of sound waves. *A whisper has low volume, and a siren has high volume.*

vol·un·tar·y mus·cle (vol′ ən tər ē mus′ əl) *noun* A muscle controlled by thought, not a reflex. See also *involuntary muscle. The skeletal muscles are voluntary muscles.*

vul·ca·ni·za·tion (vul kə nī zā′ shən) *noun* The process of treating rubber or a similar synthetic material chemically to give it useful properties, such as strength or temperature stability.

wan·ing cres·cent (wān′ iŋ kres′ ənt) *noun* The Moon phase when less than half of the left side of the Moon is visible, between the last quarter and the new Moon.

waning gib·bous (wān′ iŋ gib′ əs) *noun*
The Moon phase when more than half of the left side of the Moon is visible, between the full Moon and the last quarter.

warm-blood·ed (wôrm′ blud əd) *adjective*
Describes animals that have a fairly constant body temperature unaffected by the surrounding temperature. See also *endothermic*.

warm front (wôrm frunt) *noun*
The leading edge of a moving mass of warm air.

wa·ter (wô′ tər) *noun* A colorless, usually liquid, material made of hydrogen and oxygen. **Formula: H$_2$O**

water cy·cle (wô′ tər sī′ kəl) *noun*
The continuous circulation of water—through evaporation, condensation, and precipitation— between Earth's surface and the atmosphere.

water spout (wô′ tər spout) *noun*
A tornado that forms over water.

water ta·ble (wô′ tər tā′ bəl) *noun*
The top of the zone of rock, soil, or sediments that is saturated by groundwater.

water va·por (wô′ tər vā′ pər) *noun*
The gaseous state of water, produced when water evaporates.

watt (wot) *noun* A unit for measuring electrical power, as in the amount used by an appliance such as a refrigerator.

wave (wāv) *noun* 1. A moving ridge or swell on the surface of water, especially the ocean. *The surfer rode the wave nearly in to the shore.* 2. A vibration of air, as in a sound wave. *Sound waves are sensed by the ear.*

wave height (wāv hīt) *noun*
The distance from a wave crest to the previous trough, or low point.

wave·length (wāv′ leŋth) *noun*
1. The distance between two points of the same phase in consecutive cycles of a periodic wave. 2. The horizontal distance between two successive wave crests.

wax·ing cres·cent (waks′ iŋ kres′ ənt) *noun* The Moon phase when less than half of the right side of the Moon can be seen, between the new Moon and the first quarter.

waxing gib·bous (waks′ iŋ gib′ əs) *noun*
The Moon phase when more than half of the right side of the Moon can be seen, between the first quarter and the full Moon.

weath·er (weth′ ər) *noun* The state of the atmosphere at a given time and place. *Weather affects what we do, how we dress, and sometimes how we feel.*

weath·er·ing (weth′ ər iŋ) *noun*
The action of the wind, water, temperature changes, and other factors in breaking down rocks into smaller pieces.

a	cat	e	net	îr	gear	u	cup	u̇	look, pull	*th*	**this**	ə	alive,
ā	day, lake	ē	seed	o	hot	ū	fuse	oi	soil	hw	**wheel**		comet,
ä	father	i	fit	ō	cold	ûr	fur, bird	ou	out	zh	measure		acid, atom,
âr	dare	ī	pine	ô	paw	ü	tool, rule	th	thin	ŋ	wing		focus

weather map
(we*th*′ ər map) *noun*
A map that uses symbols
to show an area's
weather conditions.

weather sat·el·lite (we*th*′ ər sat′ ə līt)
noun A satellite that orbits over a certain
portion of Earth and takes pictures of
clouds to be used in making weather
maps and forecasts.

wedge (wej) *noun* A simple machine, such
as a chisel or needle,
with one or two sloping
sides that uses force
applied on its wide end
to drive the narrow
end into an object and
split it.

weigh (wā) *verb* To measure the pull of
gravity on an object.

weight (wāt) *noun* 1. The pull of gravity
on an object. 2. How heavy something is.

well (wel) *noun* A deep hole or shaft in the
earth through which water, oil, or natural
gas is obtained. *A water well is dug or drilled
below the water table.*

wes·ter·ly (wes′ tər lē) *noun* A wind from
the west. *Prevailing westerlies carry most of
the weather we experience every day.*

wet cell bat·tery (wet sel bat′ ə rē)
noun A device that produces direct
current by chemical reaction.

wet·land (wet′ lənd) *noun* Land area such
as a tidal flat, swamp, or marsh that
contains much moisture.

wet-mount slide (wet′ mount slīd) *noun*
A glass slide holding a specimen
suspended in a drop of liquid (such as
water) for microscopic study; also called a
wet mount.

wheel and axle (hwēl ənd ak′ səl) *noun*
A simple machine made up of a handle or
axle attached to the center of a larger
wheel.

white blood cell (hwīt blud sel) *noun*
A colorless blood cell, produced in bone
marrow, that helps the body fight
infections.

white dwarf (hwīt dwôrf) *noun*
The stage in the life cycle of an average
star in which it collapses and begins
to cool.

white light (hwīt līt) *noun* The mixture
of wavelengths of various colors that
appears colorless, like sunlight.

whorl (hwôrl) *noun* 1. An arrangement of
similar parts in a circle around a point on
an axis, such as leaves around a point on a
stem. 2. One of the turns of a univalve
shell. 3. The circular pattern of a
fingerprint.

wind (wind) *noun*
Moving air.

wind vane (wind vān)
noun An instrument that
shows the direction the
wind is coming from.

wing (wiŋ) *noun* One of (usually) a pair of
structures on animals such as birds, bats,
and insects, or on aircraft, that aid in
flying. *The shape of a bird's wing may give
clues about its flight style.*

work (wûrk) *noun* The result of a force
moving an object over a distance.

world o·cean (wûrld ō′ shən) *noun*
The continuous body of salt water that
encircles Earth and covers almost three-
fourths of Earth's surface.

X and Y chro·mo·somes (eks ənd wī krō′ mə sōmz) *noun, plural* Chromosomes that determine the gender of individuals in many species, including humans.

xy·lem (zī′ ləm) *noun* The tissue in a plant that carries water and minerals to the leaves. See also *phloem. Xylem usually carries material upward in a plant.*

ze·nith (zē′ nith) *noun* 1. The point in the sky that is directly overhead. 2. The highest point reached in the sky by a celestial body.

zo·di·ac (zō′ dē ak) *noun* The imaginary belt across the sky of the path the Sun seems to travel through, divided into 12 parts named for 12 different constellations.

zo·o·plank·ton (zō ə plaŋk′ tən) *noun* Tiny floating animal life on a body of water. See also *phytoplankton.*

zy·gote (zī′ gōt) *noun* A cell formed by the union of two gametes.

yeast (yēst) *noun* A simple fungus used to make bread dough rise.

a	cat	e	net	îr	gear	u	cup	u̇	look, pull	*th*	**this**	ə	alive,
ā	day, lake	ē	seed	o	hot	ū	fuse	oi	soil		hw	**wheel**		comet,
ä	father	i	fit	ō	cold	ûr	fur, bird	ou	**out**	zh	measure		acid, atom,
âr	dare	ī	pine	ô	paw	ü	tool, rule	th	**thin**	ŋ	wing		focus